FOREST DECLINE CONCEPTS

Edited by

Paul D. Manion
State University of New York
Syracuse

Denis Lachance
Forestry Canada
Laurentian Forestry Centre
Sainte-Foy, Québec

APS PRESS
The American Phytopathological Society
St. Paul, Minnesota

Cover photograph of maple decline stand in Allegany State Park near Salamanca, New York, by Paul D. Manion

This publication is based, in part, on presentations from a symposium entitled "Forest Decline Concepts" held in conjunction with the joint meeting of the American Phytopathological Society and the Canadian Phytopathological Society on August 7, 1990, in Grand Rapids, Michigan. This book has been reproduced directly from computer-generated copy submitted in final form to APS Press by the editors of the volume. No editing or proofreading has been done by the Press.

Reference in this publication to a trademark, proprietary product, or company name by personnel of the U.S. Department of Agriculture or anyone else is intended for explicit description only and does not imply approval or recommendation to the exclusion of others that may be suitable.

Library of Congress Catalog Card Number: 92-73742
International Standard Book Number: 0-89054-143-4

© 1992 by The American Phytopathological Society
Second printing, 1993

All rights reserved.
No part of this book may be reproduced in any form, including photocopy, microfilm, information storage and retrieval system, computer database, or software, or by any means, including electronic or mechanical, without written permission from the publisher.

Copyright is not claimed in any portion of this work written by U.S. government employees as part of their official duties.

Printed in the United States of America on acid-free paper

The American Phytopathological Society
3340 Pilot Knob Road
St. Paul, Minnesota 55121-2097, USA

TABLE OF CONTENTS

1   FOREWORD
P. D. Manion and D. Lachance

3   A HOST--STRESS--SAPROGEN MODEL FOR FOREST DIEBACK-DECLINE DISEASES
D. R. Houston

26   A NATURAL DIEBACK THEORY, COHORT SENESCENCE AS AN ALTERNATIVE TO THE DECLINE DISEASE THEORY
D. Mueller-Dombois

38   CLIMATIC PERTURBATION AS A GENERAL MECHANISM OF FOREST DIEBACK
A. N. D. Auclair, R. C. Worrest, D. Lachance, and H. C. Martin

59   THE GERMAN FOREST DECLINE SITUATION: A COMPLEX DISEASE OR A COMPLEX OF DISEASES
O. Kandler

85   A CLOSER LOOK AT FOREST DECLINE: A NEED FOR MORE ACCURATE DIAGNOSTICS
J. M. Skelly

108   ALASKA YELLOW-CEDAR DECLINE: DISTRIBUTION, EPIDEMIOLOGY, AND ETIOLOGY
P. E. Hennon, C. G. Shaw III, and E. M. Hansen

123   SUGAR MAPLE DECLINES - CAUSES, EFFECTS, AND RECOMMENDATIONS
D. C. Allen, E. Bauce, and C. J. Barnett

137 A QUANTITATIVE TREE CROWN RATING SYSTEM FOR DECIDUOUS FOREST HEALTH SURVEYS: SOME RESULTS FOR ONTARIO
D. McLaughlin, W. Gizyn, W. McIlveen, and C. Kinch

155 PATTERN AND PROCESS OF MAPLE DIEBACK IN SOUTHERN QUEBEC (CANADA)
G. Daoust, C. Ansseau, A. Thériault, and R. van Hulst

168 DEVELOPMENT OF A HAZARD RATING SYSTEM FOR DECLINE IN SOUTHERN BOTTOMLAND OAKS
V. Ammon, T. E. Nebeker, C. R. Boyle, F. I. McCracken, and J. D. Solomon

181 FOREST DECLINE CONCEPTS: AN OVERVIEW
P. D. Manion and D. Lachance

191 LITERATURE CITED

234 INDEX

ACKNOWLEDGEMENTS

We acknowledge the editorial assistance of Drs. Patrick Fenn and Donald J. Leopold. Additional editorial suggestions on individual chapters were provided by Drs. John D. Castello, David H. Griffin, Charles A. S. Hall, and Dudley J. Raynal.

We are especially grateful to Mrs. Nadia Iwachiw for typing and page layout of the manuscripts.

Travel funds for Drs. Otto Kandler and Dieter Mueller-Dombois were provided by the Canadian Phytopathological Society and the American Phytopathological Society.

FOREWORD

Paul D. Manion and Denis Lachance

State University of New York, College of Environmental Science and Forestry, Syracuse, NY 13210 and Forestry Canada, Laurentian Forestry Centre, Sainte Foy, Quebec, Canada G1V 4C7

Forests are complex dynamic communities of living and dead trees interacting among themselves and with an array of microbes, pests, environmental, human and other factors to continuously shape and reshape the community over time. Death of trees is as inevitable as birth and growth to the vitality of the forest, but when death interferes with our financial or emotional expectations, we consider it abnormal and look for a simple explanation.

"Wise men" provided simple "wrath of the gods" explanations for primitive people. The germ theory of the late 1800s, with all its contributions to modern medicine and health, has been the foundation for simplified explanations for many of the ailments of people and crops. The germ theory also has been the foundation for the development of a wealth of information on health problems in trees.

Although problems of deteriorating health and death of trees may be extremely puzzling to most people, specialists in forest pathology can usually provide a reasonable explanation. Some problems are readily diagnosed by unique signs of the pathogen. Host symptoms are often rather general but may be helpful in narrowing down where to look for the pathogen. Some problems require culturing or specific time consuming laboratory isolation and characterization techniques. Problems caused by non biotic factors have been described but may be difficult to diagnose unless one has a detailed history of the environmental and site conditions. The point to recognize is that professionals have developed the tools and information to explain many problems of trees in fairly "simple terms".

The 1980s were a decade of heightened concern for forest decline. When cursory examination did not reveal simple explanations for specific forest problems, acid rain and other environmental pollutants were hypothesized and readily accepted as a component of complex region wide forest problems.

Forest Decline Concepts attempts to provide a foundation of diverse ideas and approaches to explain some problems of trees that do not lend themselves to "simple" interpretations.

A forest decline discussion session at the 1988 annual meeting of the American Phytopathological Society in San Diego, California and a joint symposium of the American and Canadian Phytopathological Societies at their 1990 annual meeting in Grand Rapids, Michigan set the stage for this volume. Participants in these two sessions and others were invited to contribute papers to Forest Decline Concepts.

As organizers of the Forest Decline Concepts Symposium and editors of this volume, we provide a general overview and express some of our opinions in the chapter titled Forest Decline Concepts: An Overview at the end of this volume

We gave minimal instructions to the authors so as to provide maximum flexibility for expression of their views. This volume does not provide a unified concept of forest decline nor does it define the role of pollutants in decline, but by combining the key points of the various authors, a useful conceptual foundation on forest declines and a current understanding of the forest decline environmental issues can be derived.

A HOST--STRESS--SAPROGEN MODEL FOR FOREST DIEBACK-DECLINE DISEASES

David R. Houston

USDA Forest Service
Northeastern Forest Experiment Station
51 Mill Pond Road
Hamden CT 06514

The epithets "dieback," "decline" and "blight" given to many past and current tree disease problems describe either the most significant feature or the general nature of the particular disease syndrome. These terms neither identify, nor even suggest, causal relationships. Indeed, in most cases when these names were assigned, causes were unknown. As etiologies were clarified, diseases shown to be incited by specific primary causal pathogens were sometimes renamed, for example, that portion of the ash dieback complex attributable to mycoplasma-like organism (MLO) infection is now termed ash yellows, and live oak decline, now known to be incited by Ceratocystis fagacearum (Bretz)Hunt, probably will become known as live oak wilt. Most of the diseases originally named "dieback" or "decline" retain their names -- testifying to the fact that, for them, no causal pathogens capable of inciting disease in healthy trees (primary pathogen) exist. For these diseases, death of trees or tree parts is principally a consequence of invasions by organisms variously described as secondary, weak, opportunistic, facultative, or saprogenic. Such organisms are unable to incite disease in healthy, unstressed trees. From the 1940's through the 1960's, a group of problems including birch dieback, ash dieback, maple dieback, oak decline(s), maple decline(s), sweet gum blight, and maple blight appeared in hardwood forests of eastern North America. Similar problems also occurred in some conifer species, notably pole blight of western white pine (Pinus monticola Dougl.), decline of ponderosa pine (P. ponderosa Laws.) in the San Bernardino and San Gabriel Mountains of California, and littleleaf disease of shortleaf (P. echinata Mill.) and loblolly (P. taeda L.) pines in the Piedmont region of southeastern United States.

Comparing the results of research conducted on this seemingly diverse array of problems reveals a common complex-causal relationship. Preceding the onset of symptoms for each of these diseases were episodes of environmental stress, especially: unusually severe or protracted periods of water shortage or temperature increase; periods of extreme winter cold, sometimes following periods of unusual warmth; early or late frosts; and outbreaks of defoliating or sucking insects, singly or in concert. And then, before their death the buds or twigs, or stems or roots of stressed trees were invaded and killed by one or more secondary pathogenic organisms.

The chronology of environmental stress and subsequent attack by secondary organisms was first demonstrated in a series of classic studies by Bier (59-63) who showed that in excised branches, a number of saprogenic canker fungi were able to colonize bark only after its moisture content was lowered to about 80 percent of what it was when saturated.

The implication of Bier's findings and the observations of many researchers made during investigations of numerous stress-associated diseases during the 1950's and 1960's and earlier (27, 37, 42, 43, 50, 58, 84, 93, 118, 186, 192, 221, 318, 362, 372, 381, 424, 428, 442, 464), together with my own experiences with a decline of sugar maple in Wisconsin (216), led to development in the late 1960's of a working hypothesis or 'model' for these dieback-decline diseases (205, 207). This was expressed later more explicitly (209) as:

1. Healthy trees + stress → Altered trees (tissues) (dieback begins)

2. Altered trees + more stress → Trees (tissues) altered further (dieback continues)

3. Severely altered + more stress → Tree (tissues) altered further
   trees (tissues)   (dieback continues)
-
-
-

n. Severely altered + organisms of → Trees (tissues) invaded.
   trees (tissues)   secondary action   (Trees lose ability to respond to improved conditions, decline, and perhaps die.)

In these statements, the numbers refer to sequential episodes of stress events and host response; "n" indicates that at some point in time or degree of host change, organisms of secondary action are able to invade altered tissues successfully. Arrows are to be read as "leads to". A graphic conceptualization of this model appears in Figure 1.

The statements of this hypothesis can be construed as summaries of several important relationships:

i) Dieback of trees or tissues often results from the effects of the stress factor(s) alone. With abatement of stress, and in the absence of significant colonization by saprogens or secondary insects, dieback often ceases and trees recover. The dieback phase can be viewed as a survival mechanism whereby the tree adjusts to its recently encountered adverse environment.

ii) Stress alone, if sufficiently severe, prolonged, or repeated, can cause continued or repeated dieback and even death. Numerous reports exist of tree mortality following either unusually severe and prolonged drought periods, or episodes of severe defoliation, especially if repeated, perhaps even in the same growing season. When drought and defoliation occur concomitantly or sequentially, mortality can be high.

iii) Usually, however, the decline phase, wherein vitality lessens and trees succumb, is the consequence of organism invasion of stress-altered tissues. Recovery from this phase, which is now less likely to occur with abatement of stress,

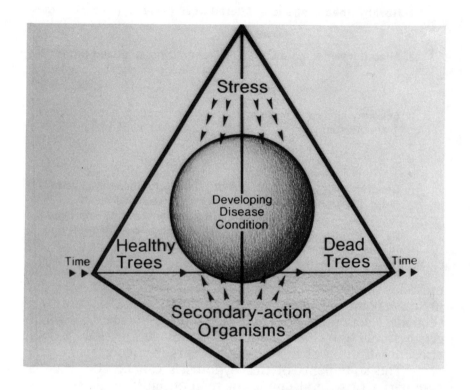

Fig. 1. A conceptual framework for the host-stress-saprogen model. Healthy trees are affected by environmental stress; tree tissues altered by that stress are invaded at some point in time by saprogens. The disease condition develops, tissues and trees die back, decline, and ultimately may die.

depends on many factors including the condition of the tree, the particular tissues invaded, the relative aggressiveness of the organisms, and the degree of invasion.

iv) Where and when the dieback phase occurs is closely related to where and when the triggering stress(es) occurs. The decline and mortality phase is related, in addition, to the temporal and spatial distributions of the organisms of secondary action.

Partitioning the developmental stages of these diseases into separate phases that encompass one or more stages of stress and host response followed by debilitating and often mortality-causing attacks of saprogenic organisms, provides a framework to aid in disease diagnosis and study. It also serves to emphasize the chronological ordering of the stress--host change--saprogen relationship. It is obvious, however, that in reality these relationships are continuums of interactions reflecting physiological and morphological host responses to a variable suite of shifting environmental stresses exacerbated by invasion of an opportunistic organism complex.

The chronological "positioning" of attacks by saprogens secondary to stress events, together with their often great diversity in number and kind, has led many pathologists to discount saprogens as important components of dieback-decline diseases. To accept saprogens as important components of causal complexes in decline diseases may require reexamining concepts of i) host predisposition and ii) the effects of stress relative to the role and consequence of secondary pathogens.

PREDISPOSITION

In the context of plant or animal disease, the term predisposition means "to bring about susceptibility to infection" (Websters Unabridged). Many organisms are able to invade tissues successfully only after those tissues have been altered or predisposed in some way. Host--stress--saprogen interactions of dieback-decline diseases thus provide some of the clearest examples of predisposition in the phytopathological sense of the term. Because stressed trees, in the absence of organism attack, usually recover (regain their former state of health, if not stature) once stress abates, a most critical stage in the development of diebacks-declines (in terms of actual damage and mortality) is the point when invasion of vital tissues by secondary-action organisms is sufficient to impair recovery.

STRESS AND PREDISPOSITION

Each of the major stress factors associated with dieback-decline diseases alters tree tissues in different ways. In many cases, neither the biochemical or physiological changes in trees or their tissues wrought by stress events, nor the mechanisms responsible for lowered host resistance are known. The fact that the same saprogen follows different invasion patterns in hosts subjected to different stresses, suggests that the nature, timing, and magnitude of stress-induced changes not only determines the nature and degree to which tissues are altered, but also, **when** they become susceptible, to **which** organisms and with **what** consequences.

In one of the few studies to systematically investigate stress-induced changes in host resistance in intact plants under controlled conditions, Schoeneweiss and coworkers (104, 399-401, 502) compared the effects of drought, freezing, and defoliation on the susceptibility of several woody ornamental species (especially Betula alba L.) to canker and dieback fungi (especially Botryosphaeria dothidea (Moug.:Fr.)Ces.&De Not). For each of these stresses, a threshold existed that had to be reached or exceeded before the saprogens, previously introduced to wounds, could invade stem tissues. Thus, water potentials in stems had to drop to -1.2 to -1.3 MPa and remain at this level for 3 to 5 days before tissues became susceptible. And, colonization increased with further decreases in water potential. Under drought stress, colonization of bark and wood was much more extensive near the cambium than in older wood. Similarly, freezing temperatures of between -20 and -30º C were required before stems of plants partially cold-hardened became susceptible. In contrast to the effects of drought stress, freezing predisposed older wood tissues near the pith to colonization. Plants remained predisposed to B. dothidea for up to 9 days after thawing (501). And, the extent of colonization increased with decreasing temperatures.

The foregoing studies examined how stresses predisposed tissues to organisms introduced to wounds. More recently, Pusey (367) examined the influence of water stress on lenticellular infection and disease development by B. dothidea in unwounded peach seedlings. In contrast to studies employing wound-inoculations (104), water stress at the time when nonwounded lenticels were inoculated (April, June) did not affect disease development, But water stress imposed 2 to 6 months after inoculation (August-October) resulted in significant increases in incidence and severity of lesions. These results illustrate the characteristic ability of saprogenic organisms to exist as saprophytes until adverse environmental conditions predispose adjacent living tissue.

Defoliation stress, when prolonged by repeated removal of all leaves for 4 weeks or more (an event unlikely in nature), predisposed bark and stem tissues to colonization by B. dothidea previously introduced to stem wounds (104). The patterns of tissue colonization were similar to those following drought stress suggesting that severe, prolonged defoliation and drought affect tree tissues in similar ways. This suggestion is supported by results from other studies. For example, changes in root tissue carbohydrates and amino acids of defoliated or droughted sugar maples and oaks are similar (348, 350, 351, 488), and both of these stresses predispose roots of eastern broadleaved trees to colonization by Armillaria sp. (442, 498). Studies of the relationship between defoliation and susceptibility of broadleaved trees to Armillaria spp. are among the few that have attempted to relate stress-induced chemical changes in hosts to the requirements of saprogenic organisms. Defoliation stress sufficient to trigger refoliation resulted in a number of significant biochemical changes. Starch reserves were rapidly hydrolyzed (488, 499) to soluble sugars, especially glucose and fructose (349, 350, 488); certain amino acids, especially asparagine and alanine increased (351, 488). Activities of some fungus-lysing bark enzymes, β-1,3 glucanase and chitinase, were reduced (489, 490). These changes are directly relevant to attack of defoliation-stressed trees by Armillaria because: i) lowered levels of hyphae-lysing enzymes allow the fungus to breach bark barriers and reach subcortical regions where ii) abnormally high quantities of glucose, the energy source most favored by the fungus, as well as growth-enhancing quantities of asparagine and alanine are available. And iii), in the presence of abundant glucose, phenols such as gallic acid that normally are inhibitory to Armillaria, are rendered ineffective, indeed are readily metabolized and become growth stimulating (493-495).

Such studies as these have revealed the existence of intimate relationships between the effects of particular kinds of stresses and specific organisms of secondary-action.

SECONDARY-ACTION ORGANISMS
(SAPROGENS, SECONDARY-INSECTS)

Secondary-action organisms include a wide variety of fungi and insects that can kill fine roots, buds and fine twigs, or bark and cambium of branches, stems, and roots. While invasion by many of these organisms can deliver the coup de grace following a given host/stress episode, certain ones have achieved notoriety because of their repeated association with one or more major decline diseases. Some of these organisms are specific to a single host genus or even species, others have a broad host range. Most significant among them are root-girdling fungi of the genus Armillaria and stem-girdling insects of the genus Agrilus, although many other bark-attacking fungi and insects cause major losses in particular situations.

In general, the saprogens associated with important forest dieback-declines:

i) are ubiquitous inhabitants of natural forest ecosystems whose evolved roles, in the absence of major external stress events, are ecologically beneficial.

ii) are unable to succeed in living tissues not previously predisposed by stress.

iii) affect stressed trees principally by invading and killing meristematic regions of roots, stems, twigs, or buds.

In natural forests, trees crowded and shaded often die -- losers in an intense struggle for light, moisture, and nutrients. Such normal attrition, frequently viewed as a simple energy relationship, is accepted and expected as a natural consequence of forest stand development. In similar ways, shaded lower crown branches of vigorous trees die and fall. Yet, close examination often reveals that death of suppressed trees and shaded branches was hastened by actions of the same opportunistic saprogens implicated as killers of vigorous trees altered by sudden, major, unexpected stress events. It is only when saprogens successfully invade tissues of trees capable of surviving the rigors of natural competition that dieback-decline diseases become important.

When host plants are introduced to new environments, new host--stress--saprogen combinations may arise. Similar consequences may result when biotic stress agents are introduced inadvertently to new ecosystems. Saprogens important in dieback-decline diseases in urban environments are often of little consequence in natural forest systems, and vice versa, because of different host-stress relationships or conditions regulating their saprophytic existence. Thus, while dieback-canker fungi such as B. dothidea, Nectria cinnabarina (Tode:Fr.)Fr., Endothia gyrosa (Schwein.:Fr.)Fr., and Leucostoma kunzei (Fr.:Fr.)Munk are rarely damaging in forest situations, they are of major importance in urban or orchard situations where combinations of water shortages and pruning wounds are common. By contrast, the root fungus, Armillaria, the principal saprogen of eastern hardwood forests, is rarely a problem in urban settings that provide little opportunity for its saprophytic existence in dead roots, stumps, or other large woody substrates.

Secondary-action organisms thus often "complete" the dieback-decline disease that began when trees were altered by a stress factor. Although these organisms are unable to successfully invade healthy trees, their presence or absence as colonizers of stress-altered tissues is often the critical factor determining whether stressed trees die or recover.

FOREST RELATIONSHIPS

The degree to which trees are altered by stress is controlled in part by their genetics and in part by the conditions of their immediate environment. Dominant overstory trees present in natural forests are those whose genotypes and spatial positioning have allowed them to survive the stresses of stand development. In many of the dieback-decline diseases investigated to date, however, the older and larger trees are the most significantly affected. Yet when subjected to severe abiotic or biotic stress factors, trees may be predisposed and attacked, often lethally, regardless of their age, size, genotypes, or soil-site situations. Typically, these diseases occur over wide geographic areas--triggered by stress factors such as severe drought, defoliation, or nutrient imbalances that can transcend individual genotypes or local environments.

In mixed forest stands, dieback-decline of only one species or species group sometimes occurs. This is readily explained in situations where host-specific biotic stress agents are involved. In cases where general, abiotic stress factors are involved, decline of but a single species can sometimes be attributed to different host-specific responses to abiotic stress (because of different phenologies, architecture, avoidance mechanisms, and so on), or to the presence, for only the declining species, of saprogens capable of invading altered tissues.

SELECTED EXAMPLES OF THE HOST--STRESS--SAPROGEN INTERACTIONS ASSOCIATED WITH DIEBACK-DECLINE DISEASES

The following examples are representative of broadleaved tree declines triggered by biotic and abiotic stress factors.

More complete listings of host--stress--saprogen interactions associated with the major dieback-decline diseases of North America that have occurred in this century are given in Millers et al (319) and in Houston (211, 213).

Dieback-declines triggered by defoliation

Sugar maple decline

Several decline diseases of sugar maple (<u>Acer</u> <u>saccharum</u> Marsh.) are recognized and a number have been studied intensively. While a number of stress factors, including road salt for trees adjacent to highways (202, 266), drought (191, 192), and root freezing, during winters with no snow cover, have been associated with declines of this species, the most important stress factor affecting forest trees is insect defoliation. Significant outbreaks of sugar maple decline and mortality have followed defoliation by a variety of insects including a leafroller-webworm complex in Wisconsin (146), the saddled prominent (<u>Heterocampa</u> <u>guttivitta</u> (Wlkr.)) in New York and New England, and the forest tent caterpillar (<u>Malacosoma</u> <u>disstria</u> Hubner) in New York, New England and Canada (3). The series of intensive studies on the Wisconsin defoliation-initiated decline (then termed maple blight) demonstrated clearly for the first time some of the cause-effect relationships of a dieback-decline disease initiated by a biotic agent (146). Maple blight was triggered by a complex of defoliators, including several species of leafrollers and the maple webworm, <u>Tetralopha</u> <u>asperatella</u> (Clem.). Mortality of trees defoliated severely for up to three successive years was associated in many cases with invasion of roots and root collars by <u>Armillaria</u> sp. (216). Subsequent trials demonstrated conclusively that dieback results when defoliation is sufficiently severe to elicit refoliation or budbreak in the same season. Death of immature terminal buds or twig tissues leads to dieback of individual branches to lower nodes where latent lateral buds flush in response the next year. Studies in Wisconsin and elsewhere, however, showed that decline and mortality of both naturally and artificially defoliated trees was usually a consequence of successful invasion of roots and root collars by <u>Armillaria</u>. Stressed trees not attacked by <u>Armillaria</u> usually recovered (498). The fungus <u>Stegonsporium</u> <u>ovatum</u> (Pers.)S.J.Hughes is a common colonizer of terminal branches of trees stressed by defoliation or drought (190, 191) and appeared to hasten twig dieback (498).

The physiological and biochemical responses of trees to defoliation and the significance of these responses to subsequent attacks by <u>Armillaria</u> (mentioned earlier) can be described as the host--stress--secondary organism relationship:

1. Healthy sugar maple trees + defoliation → Sugar maples altered
-   (dieback begins)
-
-

n. Altered trees +
○ *Stegonosporium ovatum* → Twig dieback accelerated

○ *Armillaria* sp.   → Roots, root collars invaded, trees decline, die.

Oak declines

As with maple declines, the major stress factor initiating oak declines in the northeastern United States is insect defoliation. While many different insects have triggered significant losses (e.g. 442), attention by pathologists has focused on the effects of the introduced gypsy moth (<u>Lymantria</u> <u>dispar</u> L.) relative to the attacks by mortality-causing saprogens (<u>Armillaria</u> sp.), and secondary insects (<u>Agrilus</u> <u>bilineatus</u> (Web.)) (113). Most mortality of defoliated oaks results from the rapid invasion of cambium and adjacent tissues of roots and root collars by <u>Armillaria</u> or by similar invasion and girdling of upper stems and branches by <u>A. bilineatus</u>, the twolined chestnut borer (117, 491, 492). In oaks, defoliation results in the same physiological responses and consequences (dieback) and biochemical changes that allow attack by <u>Armillaria</u> in sugar maple. And, the often rapid and heavy attack of stressed oaks by <u>A. bilineatus</u>, suggests that these changes either fulfill the insects' nutritional requirements, or in some other way lower host resistance to their attack.

Ash dieback triggered by ash rust defoliation

Dieback and occasional mortality of white ash (<u>Fraxinus americana</u> L.) occur when severe outbreaks of ash rust (<u>Puccinia</u> <u>sparganioides</u> Ellis & Barth.) result in defoliation-refoliation. This example of a defoliation-triggered dieback-decline disease underscores the fact that it is the event, not the agent, that is important. What matters is how the tree responds, and this depends on the severity, frequency, duration and seasonality of defoliation. Thus, the effects of severe defoliation by ash rust are the same as those following severe defoliation by insects: terminal branches die back, and foliage develops on sprouts originating from latent buds at lower nodes. Successive episodes of rust-caused defoliation result in progressive dieback and in forested areas, decline and mortality. Yet, in urban areas or where ash is growing as first generation stands in formerly open fields, and where saprogens (e.g. Armillaria sp.) adapted to attacking defoliation-altered trees are absent, even trees with severe dieback usually recover (Fig. 2). The general freedom from attack by canker saprogens of defoliated trees suggests that a short defoliation stress period of 10-14 days is not sufficient to alter bark of ash in the same way that drought stress does. These observations are supported by studies that show that in the short run, defoliated trees become hydrated, not dehydrated (448).

It is unlikely that the dieback-decline disease triggered by severe P. sparganioides defoliation will become known as ash dieback or decline. The disease, when known to be initiated by the pathogen, will continue to be called ash rust. Yet, there is little question, that based on symptoms alone, a diagnosis made a year or more following the defoliation episode, would classify the problem as a dieback-decline disease.

For each of these defoliation-triggered diseases, changes within tissues sufficient to predispose them to saprogens appear to occur only when the defoliation stress is severe enough to elicit refoliation and attendant biochemical responses. The refoliation response normally results in the flushing of terminal buds one year prematurely. Refoliated foliage is more photosynthetically efficient than the original complement but because refoliated leaves are both smaller and less numerous, total energy production is reduced (177). A "normal" consequence due, at least in part to lowered energy reserves, is a dieback in the subsequent year of twigs unsupported by viable terminal buds; and the development of foliage on sprouts elaborated from latent lateral buds. This refoliation-dieback response can be viewed as a survival mechanism wherein the stressed tree adjusts its altered energy balance to accommodate dormant season respiration and spring foliage production requirements.

The ability to rapidly convert immobile energy reserves to forms readily transported for use in producing new foliage and for establishing barriers to separate dead and dying tissues

Fig. 2. Defoliation-triggered ash dieback. (a) 1984, two white ash trees after 2-3 successive years of defoliation by ash rust that ended in 1982; (b) 1986, one tree was removed; recovery of the remaining tree has begun as new sprout-origin foliage develops; (c) 1988, recovery continues, with new foliage on new branches, dead wood has been removed.

from those that are living, appears to be a long-evolved defense system common to angiosperms subjected to defoliation stress. Ironically, these same physiological responses permit or trigger the invasion by organisms whose own long-evolved systems enable them to capitalize on the stress-triggered alterations.

Dieback-decline diseases initiated by drought

Oak decline

Drought appears to be the most important stress factor associated with declines of oak in the southern and southeastern United States and has been implicated in significant decline episodes in other regions as well (319). Many examples have been reported (e.g. 36, 46, 147, 206, 221, 242, 272, 309, 444, 449, 454, 455, 468). Typical of these is the decline and mortality of Quercus rubra L. that occurred on the Nantahala National Forest in North Carolina in the late 1970's (454). Growth losses in affected trees were related to water deficits from 1973 to 1978. Another extensive decline of several southern oak species (Q. phellos L., Q. laurifolia Michx., Q. nigra L., and Q. falcata Michx.) in South Carolina in 1980 and 1981 also was initiated by drought (455). In the early 1980's, a general decline of many oak species occurred over wide areas of the south. While the cause of this problem has not been determined fully, it seems likely that water shortages were involved as mortality tended to be highest on the thinnest soils and on ridgetops and southwest exposures (444, 445). A recent dendrochronological study by Tainter et. al. (456) suggests that the trees that declined or died belonged to a component of the oak population that had been weakened by a series of drought episodes in the 1950's.

Armillaria sp. and A. bilineatus almost invariably have been associated with mortality of drought-stressed oaks. Because drought and defoliation trigger similar biochemical responses (351) and often occur simultaneously or sequentially over large geographic areas, (e.g. 92, 116, 123, 164, 206, 242, 280, 324, 368, 386, 433) the existence of numerous ubiquitous saprogens and secondary insects able to capitalize on these changes can result in massive tree mortality. Hypoxylon atropunctatum (Schwein.:Fr.)Cooke, a naturally occurring bark saprogen (45) was common on declining and dead oaks in S. Carolina (455). The role of this organism in the decline of drought-stressed oaks is not yet entirely clear.

Dieback-decline of ash triggered by drought

Periodic episodes of dieback and decline of F. americana and sometimes of F. pennsylvanica Marsh. have occurred in the northeastern United States and in eastern Canada (195, 361). Because outbreaks have been associated with periods of low rainfall during each decade from 1920 to the 1970's, and were especially severe during the landmark droughts of the 1930's and 1950's (195, 297, 381, 425, 459), ash dieback has been considered a drought-triggered disease. Research has shown that two saprogenic fungi (Cytophoma pruinosa (Fr.:Fr.)Höhn. and a Fusicoccum sp.) common to shaded, naturally senescing lower branches of healthy ash, cause girdling cankers of upper crown branches and main stems of water-stressed plants (297, 381, 424). Other studies (210) showed that the bark moisture content of ash saplings regulates the incidence and severity of attack by these organisms as it does in other saprogen-canker diseases (25, 63, 401).

Generally, early outbreaks of ash dieback abated with abatement of the drought periods (297, 381). In recent times, however, ash trees growing in areas now known to harbor trees infected with viruses or MLO's (193-196, 269, 393) have continued to decline even after restoration of normal rainfall patterns. Recent and ongoing studies have revealed that declining white ash infected with MLO occur commonly in New York, New England and several other adjacent states and Canadian provinces (300). Although MLO infection interferes with stomatal opening (301), the role, if any, of these organisms is not yet clear in previous episodes of dieback-decline presumed to have resulted from drought-canker saprogen associations.

Dieback-decline diseases triggered by sucking insects

A number of sucking insects including aphids, adelgids, spittlebugs and scales have been implicated as initiators of tree dieback-decline. Often, their feeding actions predispose tissues to bark cankering saprogens. The scale, Asterolecanium sp., was found to be omnipresent on twigs of Quercus prinus L. and was suspected of creating infection courts for a twig-girdling species of Botryodiplodia (396). In another study, dieback of Fagus grandifolia Ehrh. was associated with heavy infestation by the oystershell scale, Lepidosaphes ulmi (L.)(109). DeGroot (109) was unable to fulfill Koch's postulates with species of Phomopsis, Fusicoccum, Asterosporium, and Pyrenochaeta, the fungi isolated most frequently from heavily infested dying twigs. These results might be expected with many saprogens if inoculation trials were performed with non-stressed plants. Diebacks of Tilia americana L. and A. saccharum have followed infestation by the introduced basswood thrips (Thrips calcaratus Uzel) and the pear thrips (Taeniothrips inconsequens (Uzel)), respectively. More recently, observations in Pennsylvania have suggested that young leaves of sugar maple injured by T. inconsequens may be infected and killed by anthracnose fungi (443).

Beech bark disease

An excellent example of a decline disease triggered by a sucking insect is beech bark disease. In this disease of Fagus grandifolia Ehrh. and F. sylvatica L., bark infested and altered by the beech scale, Cryptococcus fagisuga Lindinger, is invaded and killed by several species of Nectria (97, 118, 218, 279).

The symptoms of beech bark disease differ from those triggered by oystershell scale in that little branch dieback occurs. This is probably because the site of action of C. fagisuga is primarily on main stems rather than on small branches. Heavy scale infestation on main stems does result in death of some outermost living bark cells, but it is not until extensive areas of cambium are killed by one or more of the associated Nectria species that trees are significantly affected. Mortality of large trees in stands affected for the first time can be high. Small trees that emerge in the aftermath of heavy tree mortality or salvage operations rarely die quickly from girdling canker as did their parents (208). Rather, they gradually accumulate discreet cankers and become increasingly defective. Over time, many of the severely affected trees slow in growth, lose vigor, decline, and eventually may die.

Beech bark disease serves as an excellent model of the truly predispositional role of an initiating stress agent, the intimate relationship between the effects of the stress agent and subsequent attack by several closely related saprogens, and the significant role of these saprogens in causing death of tissues and trees. In North America and in Europe, the several Nectria species associated with beech bark disease differ in their abilities to attack beech in the absence of beech scale. In North America and Europe, Nectria galligena Bres. causes perennial cankers on many hardwoods, rarely on beech; in Europe, N. coccinea (Pers.:Fr.)Fr. causes annual cankers after stresses on a variety of hardwoods, including beech; in North America, N. coccinea var. faginata Lohman, Watson, and Ayers is not known to attack beech or any other species. On scale-infested beech trees, however, each of these organisms readily and rapidly invades and kills bark. In addition, another species, N. ochroleuca (Schwein.)Berk., was the only species found associated with dead and dying scale-infested beech in several stands in western Pennsylvania and in Ontario (217). The pathogenicity of N. ochroleuca on scale-altered trees has not been demonstrated experimentally.

Although the biochemical basis for this scale--host--congeneric fungal complex interaction is unknown, it probably lies in the ability of these fungi to colonize tissues whose normally effective responses to wounding or invasion are rendered ineffective by insect-secreted substances. The bark of trees susceptible to beech scale is higher in individual amino acids and total amino acid content than that of resistant trees. And, while some changes in bark phenolics are induced by scale feeding activity, the relationship of these changes to Nectria infection is not known (496).

In terms of the host--stress--saprogen model, beech bark disease can be described:

1. Healthy beech stems + *C. fagisuga* → Bark infested, altered
•   (Outermost bark cells killed)
•
•

               N. coccinea var. *faginata* (N. Am.)   Bark killed, large trees
n. Altered bark + *N. coccinea*   (Europe) → girdled, killed (first
                      N. galligena   (N. Am., Europe (?))   epidemic encounter)
                      N. ochroleuca (?)   (N. Am.)

                                                                           Bark killed, discreet
                                                                           cankers accumulate on
                                                                           small, young trees.
                                                                           Many trees eventually
                                                                           slow in growth, gradually
                                                                           decline, die (aftermath zone).

Historic Birch Dieback as Viewed by the Stress-Saprogen Model

While diebacks and declines undoubtedly have occurred for as long as environmental stress factors have adversely affected trees, it was not until birch dieback occurred that these diseases received serious attention by forest pathologists or entomologists. This major decline of yellow birch (<u>Betula alleghaniensis</u> Britton) and, to a lesser degree, white birch (<u>B. papyrifera</u> Marsh.) occurred from the early 1930's to the late 1950's. It was first noted in Nova Scotia and spread over the next 15 to 20 years through the Canadian Maritime Provinces into Quebec and Ontario and through New England into New York (44, 361). Yellow birch stands over thousands of square miles were destroyed, with the most severe damage occurring in New Brunswick, Maine, and New Hampshire.

Intense efforts over two decades to resolve the etiology of birch dieback culminated in frustration because the results were either contradictory or inconclusive. No single primary pathogen or major stress factor was found associated with the disease as it developed in different areas or at different times. And, as the massive research efforts in Canada and the United States were winding down, the disease abated as mysteriously as it had appeared. Widely varied opinions and much conjecture still exist concerning the factors that triggered losses of millions of birch trees. Yet, as unsatisfying as their results may have been to them, the approaches that the birch dieback investigators used, and the results that they obtained, fit remarkably well into a host--stress--saprogen model.

The sudden appearance (and disappearance) of this disease suggested that an abiotic factor was intimately involved in the disease. Several different abiotic stress factors were found to be associated with the disease in different areas. Average soil temperature increases of about 1^oC that occurred over a 10-20 year period were shown to cause rootlet mortality (372). Redmond (372) showed that trees whose roots were killed by increased soil temperatures responded by forming new ones at deeper levels. Pomerleau (362) noted that birch dieback was severe where shallow soils prevented rerooting at lower depths. Yet, in much of the region where yellow birch grows, and particularly, where dieback was most severe, roots formed at slightly deeper levels would grow into soil horizons where aluminum was at toxic levels (219, 220). A reexamination of this relationship may be warranted given our current understanding of how high soil aluminum directly, or high aluminum-to-calcium ratios indirectly, affect rootlet development and mortality.

Rootlet mortality also results when soils freeze to unusually deep levels. Hepting (188) reported the concept by Pomerleau that freezing of shallow birch roots, which probably occurred in Quebec in 1932, 1938, 1942 and 1943, resulted in both direct root breakage and reduced water uptake. Interestingly, a similar hypothesis has been proposed to help explain the onset of a recent episode of sugar maple decline in southern Quebec where dieback and mortality followed periods of extreme winter cold that occurred in the absence of insulating layers of snow (personal communication, D. Lachance).

Drought as a predisposing factor was generally discounted by workers who could find no consistent pattern of association with birch dieback (93, 176). Yet, most researchers agreed that for a number of years before the onset of birch dieback, precipitation was below normal and was reflected in generally reduced growth (176). Pomerleau (362) suggested that lowering of water tables during periods of water shortage was important because dieback was consistently most severe in the more drought-prone sites. Hypotheses concerning long-term effects of drought in birch dieback are mirrored almost exactly today by researchers studying red spruce decline (e.g. 230) and oak declines in the south (e.g. 444, 454).

An intensive search disclosed no common primary insect or causal pathogen. Insect defoliation by a complex of leaf miners and skeletonizers appeared to initiate birch dieback in some areas but not in others (93). Several viruses in birch were detected (160, 161). And, because some of the disease symptoms occurred in healthy trees inoculated by grafting (56), there has been speculation, not yet demonstrated, that viruses might predispose trees to other agents. Because of our current and emerging abilities to detect and transmit tree viruses, and due to their presence in other species affected by decline, these agents, and others such as MLO's (that were not even recognized when the disease was studied) deserve further study. The "spreading" pattern of occurrence of birch dieback and the absence of one universally present and obvious climatic relationships prompted several researchers to suggest that a still undetected biological agent might be involved (93, 176).

Many saprogenic organisms were found to attack birch trees subjected to one or more of these stress factors. The bronze birch borer, <u>Agrilus anxius</u> Gory (43) was associated almost invariably with advanced stages of dieback and was considered to be a "decisive factor in the eventual death of the trees" (176). Only after repeated attacks were healthy trees successfully invaded -- a situation that increased as dieback intensified and larger numbers of beetles became available. <u>Armillaria</u> sp. was found associated with declining trees in some cases but not in others (160). Many fungi were isolated from buds and twigs of declining birches, but one, <u>Diaporthe alleghaniensis</u> R.H.Arnold (<u>Phomopsis</u> sp.) proved to be pathogenic although weak on vigorous yellow birch seedlings (27, 204). This fungus caused branch dieback on seedlings whose vigor was reduced by growing in shade.

The stress factors and secondary organisms discovered in studies of birch dieback can be arranged according to the general stress-saprogen model:

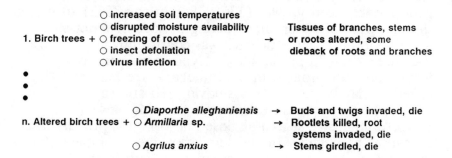

The foregoing examples are limited to conditions of broadleaved species. Many examples also exist of stress-initiated declines of conifers, including little-leaf disease of shortleaf (<u>P</u>. <u>echinata</u>) and loblolly (<u>P</u>. <u>taeda</u>) pines, pole blight of western white pine (<u>P</u>. <u>monticola</u>), and a decline of ponderosa pine (<u>P</u>. <u>ponderosa</u>) in the San Gabriel and San Bernardino Mountains near Los Angeles, California. The stress--host responses--saprogen interactions of each of these long-term research problems can be outlined as hypotheses statements of the model presented here for the hardwood dieback-declines. An example of a recent conifer "decline"

problem "viewed" in context of this model follows.

Current Red Spruce "Decline" as Viewed by the Stress-Saprogen Model

Growth declines and mortality of red spruce (Picea rubens Sarg.) in high-elevation forests of the eastern United States have received much attention in recent years. The extraordinary interest paid these conditions arose primarily from presumptions that they paralleled recent reported episodes of tree decline (principally P. abies (L.)Karst.) in western Europe triggered presumably by anthropogenic airborne pollutants (AAP), especially acidic compounds (410). Studies in North America thus far have failed to implicate AAP as a direct cause of spruce decline (e.g. 302, 357, 514), although Wilkinson (511) and DeHayes (110) showed that acid deposition results in cuticular erosion and winter dehardening. Should research eventually implicate AAP as stress(es) that can trigger decline of growth and death of spruce, it is likely that the relationship will be an indirect one. One hypothesis currently being tested is that increased acidification resulting from deposition of acid materials on thin, highly organic, poorly buffered, upper elevation soils results in the solubilization of aluminum which, at high levels, inhibits uptake of critical cations, especially calcium (422). As a consequence, growth of roots and stems is reduced, and defense mechanisms are impaired. Other studies on declining red spruce indicate that decreases in fine root vitality precede or occur coincidentally with deteriorating crowns (497).

Recent surveys, however, have revealed that many other factors also might be involved. Drought has been correlated with sharp reductions in growth of red spruce (230), and periodic episodes of dieback appear to be correlated with years in which unusual periods of extreme warming in February-March were followed by return to normal below freezing temperatures (134, 228, Personal communication, P.M. Wargo). And, many declining trees show damage to crown and roots caused by wind (168, 377). Other studies suggest that reduction in red spruce growth may be part of normal stand dynamics of released red spruce (475). LeBlanc (274) and LeBlanc and Raynal (275), however, relate growth reduction to crown injury and decreased apical growth suggesting the action of an atmospheric stress.

Numerous saprogenic organisms are associated with declining red spruce including the root fungi <u>Armillaria</u> sp., <u>Scytinostroma galactinum</u> (Fr.)Donk, <u>Perenniporia subacida</u>(Peck)Donk (87, 378), the canker fungus, <u>L</u>. <u>kunzei</u> (317); and bark borers, <u>Dendroctonus rufipennis</u> (Kirby) (307).

In terms of the stress-saprogen model, our current understanding of red spruce "decline" can be expressed in the hypothesis:

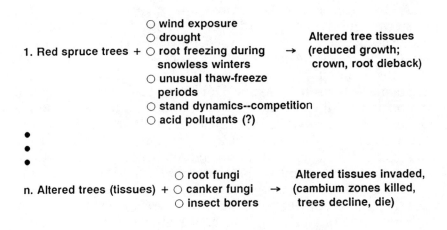

SUMMARY

Diebacks-declines are complex diseases that can result from combinations of factors--the specific mix of which may vary from area to area, or episode to episode. New host-stress-saprogen combinations will continue to occur as biotic stress agents invade new territories (e.g., gypsy moth, beech scale, hemlock adelgid, etc.), as new combinations of environmental stress events occur, or if climates change sufficiently to adversely affect tree defense mechanisms against saprogens.

Trees altered by different stress factors, and subsequently invaded by various saprogens, secondary insects or both, may exhibit essentially similar dieback-decline syndromes. While some of these diseases may be exacerbated by overall changes in climate, or increasing tree or stand maturity, most are initiated by more local, often host-specific, stress factors.

Because attack by saprogens usually determines whether stressed trees survive or succumb, organisms of secondary action are significant components of host--stress--saprogen dieback-decline models.

A NATURAL DIEBACK THEORY, COHORT SENESCENCE AS AN ALTERNATIVE TO THE DECLINE DISEASE THEORY

Dieter Mueller-Dombois
Department of Botany, University of Hawaii
3190 Maile Way
Honolulu, Hawaii 96822

When trees are dying in large numbers in an area, it is usually assumed they are dying from some kind of disease caused by a biotic agent. Phytopathological investigations, however, have established that in many such cases, other causes are involved in initiating forest decline or significant reductions in the vigor of forest stands. These other causes, as a group, have often been defined as abiotic causes. The physiological abnormalities of either biotic or abiotic disease are a reflecion of physiological damage or biochemical changes.
Whenever physiological abnormalities are clearly associated with non biotic factors (e.g. mineral deficiency), the specific symptoms are often regarded as indicative of a specific abiotic (physiological) disease. Whenever physiological abnormalities are clearly identified as caused by parasites, including insect pests, the symptoms and associated signs are regarded as indicative of a biotic disease. In this way, and in accord with Manion (291), many abnormalities in tree growth can be referred to as either biotic or abiotic (physiological) diseases.
A third group of tree disorders involving a combination of abiotic and biotic stresses have been recognized by Manion (291) as decline diseases. Because of their multifactorial origin, decline diseases are difficult to interpret.

Houston's (207, 209, 210) interpretation of decline diseases involves abiotic stresses, mostly in the form of injuries, which alter the physiology and predispose the trees to dieback and decline, caused by biotic agents. In Houston's interpretation, biotic agents are the killers, which, however, can become effective only after the trees are weakened by abiotic stresses.

Sinclair (429, 430) introduced the now well-known three-step chain reaction theory of causal factors, namely predisposing, inciting, and contributing causes. This theory has been incorporated by Manion (291) into a diagrammatic model, known as Manion's decline disease spiral. This concept of a death spiral is useful in that it brings out the continuity of the decline process. Moreover, a number of more specific causes are listed in the spiral, which in the causal hierarchy appear below the three general factors of Sinclair (429).

The list of subhierarchical and more specific causes include natural as well as anthropogenic factors. Among these a number of biotic agents are emphasized which make up the third hierarchical category, the contributing factors. Manion's concept of combining natural and anthropogenic stresses with biotic disease agents into an all inclusive model of a death spiral has a strong pathological emphasis. Therefore, in his view, decline is a form of disease. It is my intention in this paper to point out that forest decline can be distinguished as either a disease or a natural phenomenon. This, I believe, can be done if we focus not merely on the symptomology of the dying trees but also on forest decline in the context of population, community and ecosystem dynamics.

DECLINE AND DIEBACK AS A POPULATION PATTERN

I define **decline** as a stand-level or population-level phenomenon, which is manifested in a distinct loss of vigor of a forest stand, whereby many trees of the same species show leaf discoloration out-of-season and an associated decrease in diameter increment. I consider **dieback** to mean a further progression of decline, whereby crown foliage is dropped out-of-season exposing barren branches, which may lead to stag-headedness of many individual trees. Dieback in this sense can be reversible to some extent, but in most cases it ends in the death of many trees. The terms dieback and decline are so close in meaning that they can be used interchangeably.

The progression of both decline and dieback may vary from very rapid to extremely slow in individual trees. For a whole population forming a stand, the decline and dieback progression, although synchronized by definition, may be considerably prolonged by the behavior of individual trees, lasting perhaps several decades. Some individuals of the same population may survive the dieback phase and continue to grow into much bigger trees. This does not render dieback to be automatically explicable as a thinning process or a disease.

At this point, dieback is merely a structural/dynamic pattern, which I have also called **canopy dieback** in previous papers (325, 334) to emphasize the stand-level aspect of dieback. However, it resembles the decline/dieback phenomenon as seen in a population of annuals or therophytes, except that the monocarpic dieback of annuals is typically in synchrony with climatic seasonality, while canopy dieback of trees relates to polycarpic perennials losing vigor or dying out-of-season.

Another important point relating to the term canopy dieback is that only the canopy trees of a species are dying and not necessarily also its seedlings or saplings. The latter may persist in the same stand undergoing canopy dieback, or they may co-occur and persist in spatially separate sites in the same general territory like differently aged plantation stands co-exist side-by-side in a managed forest system.

In fact, dieback as a population pattern implies the recognition not only of species and individuals forming a stand and occupying a territory of various spatial dimensions, large or small, but also the recognition of individuals within a species forming a cohort, i.e. members of the same generation. Separate cohorts are formed by seedlings, saplings, mature and senescing trees growing in the same forest biome. Of course, there are possibly more recognizable age-states or life stages in trees, and there are several methods for "life-staging" individuals and groups of individuals. Such methods are part and parcel of plant population demography (144, 166, 167, 345, 426, 476, 503).

Finally, dieback as a population pattern does not mean that only one species is dying in a forest. Dieback may affect more than one canopy species in a forest at more or less the same time, although single species canopy dieback seems to be more common.

DIEBACK IN THE CONTEXT OF COMMUNITY DYNAMICS

Commercial forests, at least in Europe, are expected to show no dead or dying trees, because the foresters are checking frequently for overmature or senescing trees with their Swedish increment hammers. The trees with very low or zero diameter increments are then selectively cut in all-aged stands or the entire stand is clear-cut when it consists of an aging cohort of planted trees. Cohort forest management of Norway spruce (Picea abies (L.) Karst.) is currently the most prevalent form of commercial forestry in Europe, and it is these artificial forests that are receiving most of the attention in the European forest decline syndrome (406), the so-called "Waldsterben" (410, 411) or "novel forest disease" (264).

In contrast, dead or dying trees are a normal sight in any natural, unmanaged forest. Of course, large numbers of dead standing trees are not such a common sight unless a forest has suffered a wild fire. Forest biomes, in which wild fires are part of their natural disturbance regimes, such as the boreal forest of North America (178), often evolved two dynamically significant characteristics:

i) In areas of high fire frequencies, pine forests usually form mosaics of spatially separated even-aged stands or cohorts rather than all-aged stands. Unlike plantation stands, these natural cohort mosaics are of rather irregular shapes. They also may have spatially irregular and unequal representations of different age-states, i.e. a large spatial segment may be in juvenile life stages and another in mature and overmature or senescing life stages, while intermediate life stages are rare or absent. The forest growth cycle of such a community system can then be said to be in a non-equilibrium phase. It is often the late-mature or senescing cohort stands that attract the next wild fire because of the increased fuel loading associated with aging cohort stands. Because of the high fire frequency in such forest systems, dieback stands, as defined above, are a rare phenomenon. They can, however, be expected to be more common where fire prevention is actively pursued as a forest management policy.

ii) Fire in these forests is known also as a factor of forest rejuvenation. Their dynamic response typically is an "auto-succession", implying replacement of the fire-killed stand by the same species, usually again in the form of even-aged cohorts. From the viewpoint of canopy species diversity, this can also be called a "chronosequential monoculture".

If fire does not recur, and the critical period of high fire hazard in aging cohort stands has passed due to the invasion of hardwoods such as aspen and/or birch, successional development may proceed to spruce and fir, depending on habitat and seed availability. In such situations a full successional sequence of different dominating tree species may be realized on the same forest site, which from the viewpoint of biodiversity resembles a "chronosequential polyculture".

Both these types of forest dynamics differ very much from that found in multi-species tropical lowland rainforests, in which single treefall gaps rather than larger-scale fire disturbances are the rule (74). Single tree- or small-group break-downs result in a "chronosequential gap rotation" or small-patch mosaic system involving many ecologically different tree species as gap fillers. Such small-area community mosaics are typical for continental lowland rainforests but not necessarily also for tropical island and mountain forests.

Forest decline and dieback appears to be a natural phenomenon of community dynamics in forest biomes characterized by auto-successions. In the Hawaiian Metrosideros polymorpha Gaud. rainforest biome, we have recognized three types of dieback/recovery patterns, "replacement dieback", "displacement dieback" and "stand-reduction dieback" (326).

Replacement dieback in the Hawaiian rainforests is a form of auto-succession, although we have found some evidence in M. polymorpha for a conspecific racial shift in the successional turnover of this species (333, 446, 447). The term displacement dieback refers to a successional displacement of the dominant canopy species by other canopy species, as is typical in chronosequential polycultures. In Hawaii this occurs naturally on nutritionally rich soils from volcanic ash, where tree ferns (Cibotium spp.) are especially abundant and vigorous (83). It also occurs artificially in areas, where introduced alien tree species invade Metrosideros dieback stands in certain habitats of its wide distribution range (226). Finally, stand-reduction dieback refers to situations where a Metrosideros dieback stand recovers only in form of Metrosideros scrub, i.e. where high-stature stands give way to low-stature stands due to site deterioration for the growth and development of this native tree species (203, 329). While stand-reduction dieback may most closely resemble that of the European forest dieback, stand-reduction dieback in Hawaii can also be explained as a natural phenomenon (327, 329) as will be discussed in the next section.

DIEBACK AND ECOSYSTEM DEVELOPMENT

Terrestrial ecosystem development may start with a new geological substrate emerging from a receding glacier, or from a large denuded surface that was subjected to sheet erosion, or on a rock pile surfacing as a new volcanic mountain in the ocean. Somewhat different biotic access situations are expected with these scenarios. Biota can access a new volcanic island only by long-distance transfer, while the eroded surface or deglaciated substrate may be receiving disseminules from already existing plant communities in the proximate neighborhood.

In biogeographically isolated areas, such as islands and mountains, successful establishment of new tree populations, following long-distance transfer of naturally dispersed disseminules, can be considered a very rare event (127). Once established, a tree population, if successful, will spread locally across the available surfaces. On a volcanic mountain, the invasively spreading tree species will encounter a spectrum of edaphic conditions, from recent lava flows to older, weathered ones, including various forms of pyroclastics, such as cinder cones and ash-blanket deposits.

Biological invasion of such a habitat spectrum, which also includes short-distance changes in temperature with altitude and precipitation in relation to the prevailing winds, cannot be visualized as a uniform territorial expansion. Edaphic, topographic and climatic barriers will cause slowdowns or stagnations in the advancing tree species. However, in the absence of competing species, a generalistic colonizer may eventually succeed even in temporarily water-logged habitats.

In this process of geographic radiation, the invading tree species may be forced to genetically adapt when encountering more radical differences in substrate and physiography. This form of adaptive radiation cannot be assumed as a smooth and continuous process; instead, seedling invasion may succeed during favorable weather and climatic conditions up to a point when unfavorable conditions produce set-backs. In fact, one can consider stand-level dieback to be a natural process in the struggle of evolutionary adaptation of a tree species advancing into a new and unfamiliar environment.

The exact counterpart of this is the planting of tree species into the "wrong site", which has been proposed by Mayer (303) as the most probable cause of the silver fir (<u>Abies alba</u> Mill.) dieback in south-central Europe. A tree species, under these conditions, may grow well into maturity. At that time or sooner, weather perturbations, such as a late frost or a period of low radiation associated with prolonged rain and cloudiness followed by drought, may interact negatively with a poorly buffered soil environment. This then can result in stand-level dieback through physiological shocks. Two physiological impediments may come together here: i) the loss of energy due to the gradually increasing respiratory wood volume/leaf-area ratio with the aging of trees and, ii) cavitation or xylem embolism, a physiological dieback cause, recently proposed by Auclair (31).

Ecosystem development, unfortunately, does not terminate with the idealized climatic climax of Clements (95). Instead, at least in the island tropics of Hawaii, centuries of progressive soil enrichment with organic matter and biophilic nutrients are followed by millennia of regressive soil development in the rainforest environment as evidenced by the side-by-side development of differently aged islands. The year-round high precipitation level promotes soil acidification from organic acids, which in turn cause leaching of the cations and a decrease in pH from 6.0 to 3.5 with aging of the basaltic volcanic substrates. This leads to aluminum (320) and occasional manganese and iron toxicities in poorly drained soils (35). Immobilization of phosphorus may be another principal cause of productivity decline (485). Stand-level dieback on such nutrient limited, partially toxic, and poorly drained, aging soils was identified as "stand-reduction dieback", implying a replacement of forest by scrub vegetation.

This is a natural regression in productive capacity not unlike that, which forced indigenous people in the humid tropics into the practice of shifting cultivation. It is also analogous in mechanism to the European forest decline, now identified as caused primarily by cation deficiency (526).

A NATURAL DIEBACK THEORY BASED ON COHORT SENESCENCE

The foregoing discussion, which focused on dieback in the context of population, community and ecosystem ecology, can now be summarized into a natural dieback theory based on cohort senescence. This theory can be simplified by focusing on four principal generic causal factors as follows:

s = Simplified Forest Structure

Simplified forest structure and canopy species diversity may occur as a spatially dynamic mosaic of cohort stands dominated by the same canopy species. Such simplified structure destabilizes a forest system in two ways: i) by creating a predisposition to stand-level dieback in those cohort stands that have grown beyond their most vigorous life stage into a phase of senescence, and ii) by creation of chronosequential monocultures, which can degenerate the soil resource by one-sided nutrient extraction and the excretion of poorly degradable metabolic substances that may result in autotoxicity. Both factors led to the need for crop rotation in European agriculture.

e = Edaphically Extreme Sites

Edaphically extreme sites may support dieback susceptible species that are not yet perfectly adapted. Thus, a species may suffer from not having achieved a good evolutionary fit to its habitat. Such edaphically extreme sites may be invaded by the species under favorable climatic conditions, but growth impediments may soon be experienced by nutrient limitations or unfavorable soil moisture conditions. Typically, these are shallow soils, which are poorly buffered against climatic extremes.

p = Periodically Recurring Perturbations

Periodically recurring perturbations or fluctuating site factors may occur as weather disturbances or climatic instabilities that temporarily stress a forest stand far beyond what it experiences as normal seasonality. In contrast to catastrophic disturbances such as fires, hurricanes, or volcanic blanket deposits, which cause extensive physical destruction, the periodically recurring perturbation pulses will primarily be manifested in physiological shocks in the impacted forest. They may, however, cause premature growth stagnation if recurring frequently and like factor "e" may add to the premature senescing of cohort stands. Eventually, the "p" factor may occur as the dieback trigger in the sense of Sinclair's (429) inciting factor.

b = Biotic Agents

Biotic agents may either contribute to weakening individual trees and cohort stands by periodic attacks, similarly as the "p" factor. But, in contrast to the "p" factor, which is independent of the dynamics or physiological condition of the host trees, biotic agents usually become overpowering only in the end-phase of the tree's growth capacity. They provide the *coup de grace* and may thereby hasten the dieback. From the viewpoint of community dynamics, this may become an advantage in the recovery process following dieback particularly in biotically impoverished systems, where forest recovery depends on auto-succession. Biotic agents which become epidemic, when trees are dying from other primary causes, may also fulfill a role as precursors of decomposers, particularly in cold, wet montane forest ecosystems where fire is rare and normal decomposition processes are particularly slow.

These four factors, all of which occur naturally and independently of anthropogenic inputs, can now be fitted conveniently into Sinclair's (429) three-step chain reaction model, used also by Manion (291) for his decline disease theory:

i) Predisposing factors
- s simplified cohort structure and senescence
- e extreme edaphic and evolutionary stresses
- p pulse perturbations, i.e. periodic physiological shocks from extreme weather or from seismic disturbances

ii) Inciting factor	p	pulse-perturbation factor (same as "p" above) that triggers canopy breakdown in demographically weakened stands
iii) Contributing factor	b	biotic agents, such as pathogenic fungi or insect pests which may overpower a stand weakened by the preceding three causes, s, e, p

CONCLUSIONS: NATURAL DIEBACK VERSUS DECLINE DISEASE

The argument can certainly be made that forest decline is a disease, because in the end, trees usually are overpowered by pathogens. One may also argue that dieback is always a physiological disorder and therefore a disease.

However, aging is not a disease. Moreover, aging is not merely a calendar progression but a function also of the stress-history or demography of an organism and its genetic constitution. The same applies by extension to a cohort of individuals living together in the same community. The natural dieback theory differs from the decline disease theory by the addition of a third component to the explanation of stand-level dieback. Disease, in the above more limited sense, certainly is one component, and abiotic stresses are a second. The third is the demographic component, i.e., forest stand demography in the broadest sense. This includes the demographic behavior of the dieback susceptible species at the level of population biology, but also in the context of its community and specific ecosystem.

At the level of population biology, a tree's physiological ecology is expected to change from seedling to sapling, when most of the energy is allocated to vegetative growth. At maturity, resource allocation is divided between vegetative and reproductive effort, and beyond peak maturity when the energy-level decreases, trees, like other organisms, may lose some of their defense mechanisms or capacities for damage repair. Senescence in polycarpic trees is an almost unexplored field of physiological and genetical ecology (145, 516), and even the physiology of monocarpic senescence is not yet very well understood (463). Stand demography includes also various aspects of community dynamics, in particular the functional diversity of canopy species with respect to their successional roles. For example, a complete assemblage of pioneer, seral and climax species may occur in one forest system. In biotically impoverished systems one or the other of these tree life forms may be absent or rare, which in turn will result in the destabilization of that system. Moreover, the stand demographic component includes the historically operative stress factors, such as the effects of edaphically extreme sites and the disturbance regimes, which are an ecosystem-specific characteristic of any larger biome.

Finally, the cohort senescence theory which is interpreted as having a primarily demographic basis, is applicable also to forests whose canopy trees consist of different age classes and many species. The difference is, however, that synchronously dying cohort members may be dispersed throughout a larger stand of mixed species and/or among the more vigorous younger members of the same species. From the air, such forests may reveal a pattern of "salt-and-pepper" dieback.

In very diverse multi-species tropical rainforests, species members of the same generation may not co-exist in the same canopy. In such cases, cohort senescence cannot be implicated. However, individual tree decline and dieback may still be initiated by senescence as a predisposing factor. Another difference between canopy dieback and the typical gap dynamics of multi-species rainforests is that in tropical lowlands dead or dying trees do not remain standing for very long because of the rapid activity of decomposers in the tropical "greenhouse climate". This may be a reason why dieback trees and the etiology of individual tree death have not received much attention in the "patch dynamic" theory of ecologists (356). Instead the attention of ecologists has so far focused on the consequences of individual tree or tree group dieback (358) and not on the causes of tree mortality.

For evaluating the impact of new anthropogenic stresses, such as air pollution, climate change and biotic impoverishment, it is important to understand the natural processes of forest dynamics. Only then will it be possible to untangle the real impact of human influences on forest decline and dieback.

ACKNOWLEDGEMENTS

Much of the background research for writing this paper was made possible through NSF Grants BSR-8416178 and BSR-871004. I also wish to thank my wife, Annette Mueller-Dombois, for her continued support and assistance in word processing.

CLIMATIC PERTURBATION AS A GENERAL MECHANISM OF FOREST DIEBACK

Allan N.D. Auclair[1], Robert C. Worrest[2], Denis Lachance[3], and Hans C. Martin[4]

[1] Science and Policy Associates, Inc., Suite 400 West Tower, 1333 H St. N.W., Washington, DC 20005,

[2] Office of Research and Development (RD-682), US Environmental Protection Agency, Washington, DC 20460

[3] Laurentian Forestry Centre, Forestry Canada, Sainte-Foy, Quebec, G1V 4C7

[4] Atmospheric Environment Service, Environment Canada, Downsview, Ontario, Canada M3H 5T4

Forest pathologists have recognized since the 1940s that region-wide forest diebacks can be closely associated with shifts in climate (175, 176, 360). Evidence of this was first consolidated through broadly focused but comprehensive reviews of climate-incited diseases on North American tree species (187, 291). Progressive dieback and mortality on yellow birch (Betula alleghaniensis Britton), white ash (Fraxinus americana L.), red spruce (Picea rubens Sarg.), European silver fir (Abies alba Mill.), and Norway spruce (Picea abies (L.)Karst) have been linked to changes in climate (55, 102, 229, 273, 372, 381).

Researchers have only recently provided the historical evidence that episodes of wide-scale tree dieback and mortality have occurred under natural conditions, as in some regions of the Pacific rim (326, 332), and over periods early in this century and in the previous century (100, 157, 172, 179, 352) long before regional air pollution became a serious problem. Of particular interest because of their remoteness from population and industrial centers were extensive, early (1880s) diebacks on Alaska yellow-cedar (Chamaecyparis nootkatensis (D. Don) Spach.) over the Alexander Archipelago, Alaska (179), and on black spruce (Picea mariana (Mill.) B.S.P.) in the forest tundra of northern Quebec (352).

That changes in climate could be an immediate cause of forest dieback is supported by several simulation studies on forest response (51, 52, 54, 55, 96). Solomon (440) simulated the potential shifts in forest species composition and forest structure under scenarios of global warming; of particular interest was the sensitivity of northern hardwood forest to dieback under climatic conditions expected in the near term. His and related studies have emphasized the need to consider seriously the possibility that global warming may lead to dieback of some of the dominant tree species (122, 439).

One consequence of the failure of efforts so far to conclusively link air pollutants to forest dieback has been the development of a multitude of hypotheses (353) and an acrimonious debate on the causal mechanisms (198). This lack of resolution may be the motive for adopting a new perspective, different approaches, or for exploring the problem in a much larger context (68).

The senior author extensively re-investigated early forest diebacks (28, 29, 30, 32, 483, 484). Our objective in this paper is to develop a general theory that is consistent with historical and current observations of various forest diebacks and to formulate specific testable hypotheses on the mechanism(s) of dieback. There is a need to organize old as well as new information into a consistent, unified perception on dieback mechanisms.

METHODS AND DATA

We felt it would be particularly instructive to develop a rigorous case study of forest dieback in the Canadian and U.S. northern hardwoods. Diebacks here have been researched intensively since the spectacular dieback on birch in the 1930s and 1950s. Together with diebacks on ash, oak (Quercus), and sugar maple (Acer saccharum Marsh.), these served as the basis on which concepts and definitions of forest dieback were first developed (209, 291, 428).

The dieback of birch was especially distinctive. Numerous well-designed experiments over three decades allowed us considerable perspective in interpreting much of the early work on birch in a modern context. A further dieback on northern hardwoods is in progress making it an historical as well as a contemporary phenomenon. This has provided the added advantage of data on alternate deciduous species (sugar maple) and comprehensive information from the most recent survey and monitoring programs.

Persistent, progressive, region-wide forest crown dieback was chosen as the symptom on which to base the analysis. Among the symptoms on individual trees are: early leaf coloration and premature leaf-fall, leaf chlorosis, small leaf size, tufting of leaves, and stem dieback from branch tips inward toward the bole. These features are readily observed and were the most consistently reported of the ten common denominators among forest diebacks in North America (291). Reduced radial growth is also characteristic of the dieback. We chose not to focus on this symptom since a straightforward interpretation can be complicated by many influences on tree growth. Unlike crown dieback, radial growth decrease is not necessarily pathological. Radial growth reductions can be associated with years of prolific seedset or natural aging.

An important distinction made early in the analysis was to focus, not on forest "declines" in general, but on one specific type, namely region-wide forest crown dieback which can occur as one of the more extreme forms of forest "decline". There remains a large conceptual gap in the classification of forest "declines" and diebacks and few papers which integrate early research (1940-1970) focused on climate with the current focus on air pollution. In this study forest decline refers to a progressive reduction of productivity and health caused by a wide range of potential single factors, or interacting factors. Forest crown diebacks, when region-wide, persistent, and progressive, are perceived to be one specific type of forest decline.

Region-specific data sets on forest dieback in eastern Canada were based on annual forest pathology surveys of the Forest Insect and Disease Survey unit of Forestry Canada, (formerly the Canadian Forestry Service), government reports, and published literature (483, 484). For each tree species showing widespread and persistent (3-30 years) crown dieback, we enumerated in chronological order the location, severity, symptoms, and the assigned cause of the dieback episode. The outcome was a set of tables documenting the history of onset, development, and in some cases, the recovery of individual tree species in eastern Canada. Eastern Canada represented the northern half of the northern hardwoods formation and the area of the most severe diebacks.

The severity of crown dieback was plotted over the geographic distribution of each tree species (129). Graphs were also made of changes in the extent and severity of forest crown dieback over time; this helped us to accurately identify onset and recovery periods for climatic analysis.

An "event analysis" was made of the 1981 anomalous winter conditions. An intense warming event in mid-February was followed by deep freezing in March, coinciding with the onset of massive dieback of sugar maple in southern Quebec the following summer (268). Daily maximum and minimum air temperatures at Lennoxville, Quebec, from November 1980 through May 1981 (12) were plotted to verify the degree of thaw-freeze (390) and to interpret its physiological effect. To achieve the latter, we used the winter frost resistance limits of cold continental temperate deciduous tree species published by Larcher and Bauer (271). Summaries of the rate of winter stream discharge from the adjacent Chaudière River watershed (15) were used to verify the presence of a regional thaw sufficiently intense and prolonged to melt the snow pack; this had the implication that a thaw of such a magnitude could also initiate early sap flow and put the tree at risk of freezing injuries.

To test the universality of a climate-dieback mechanism, an event analysis of the onset year was applied to each of the major early dieback episodes in northern hardwoods and to diebacks in two other regions, namely that on western white pine (Pinus monticola Dougl.) in the Pacific Northwest (273) and on Norway spruce in Central Europe (29, 304, 373, 374).

RESULTS AND DISCUSSION

Crown Dieback - Historical Incidence

Three episodes of widespread dieback have occurred in the northern hardwoods of Eastern Canada. In 1925 (or 1927)

(360), dieback on black ash (<u>Fraxinus</u> <u>nigra</u> Marsh.) was widespread across southern Quebec, being particularly severe on wet sites and sometimes on isolated trees growing on thin soils. First symptoms may have appeared earlier (360, 361, 363) but our reconstruction indicated that 1925 was the first year of severe crown injury. Another episode occurred in 1932 (363) that was first noticed in 1935 (or 1936) on white ash (360). A further incidence of dieback on ash species appeared about 1960 and in the late 1970s to present (Fig. 1A).

Several authors reported independently the first signs of dieback on birch (<u>B</u>. <u>alleghaniensis</u> and <u>B</u>. <u>papyrifera</u> Marsh.) between 1932 and 1935, (4), the onset of massive crown dieback first occurring in 1937, 1938, and 1939, depending on region (72). In 1937, dieback on birch was particularly severe (Fig. 1b) and was observed in the Laurentian Mountains north of Montreal for the first time (361). The first episodes of extensive mortality started in southern New Brunswick (41). Dieback subsequently increased in intensity. Mortality of birch in New Brunswick was particularly severe in 1940 and 1941 (41), and extensive in Quebec between 1946-1949 (361).

The first signs of recovery on birch were observed in 1944 in New Brunswick (175) and in restricted areas of the Gaspé Peninsula in 1946 (106). Several local areas of recovery were observed the following year (1947) in southern New Brunswick (175) and by 1952 dieback had practically ceased in Nova Scotia and New Brunswick (41). Pomerleau (363) observed that unlike in the Maritimes, the post-1944 recovery did not occur in Quebec. A general recovery throughout Quebec was apparent only during and after 1950 (361).

Some resurgence of dieback occurred in Lake St. John-Saguenay-Laurentian Park region of Quebec in 1957 (281, 360) and on yellow birch (1966-1970) over the east shore of Lake Superior in Ontario. Residual pockets of dieback on white birch have persisted on the east shore of Lake Superior from 1950 to the present (G.M. Howse, personal communication 1986). Severe crown deterioration on white birch (<u>B</u>. <u>papyrifera</u>) has been apparent since 1979 along the perimeter of the Bay of Fundy in New Brunswick and Nova Scotia (288), and throughout the Northeast United States since 1984 (500).

The first reports of sugar maple dieback in Canada were from Quebec in 1932 (360) and from Ontario in 1947 (308). Since then, the incidence has been sporadic but localized in these provinces. Three aspects of the recent dieback episode on sugar maple were noteworthy. The first massive dieback was not apparent in southern Quebec until the summer of 1981. (Fig. 1C). Several questionnaires mailed to maple sugar producers in the province confirmed that south of the St. Lawrence River, the disease had been scattered and sporadic since 1978, but not extensive. North of the St. Lawrence River in Quebec (88) and in Ontario (310) crown dieback was evident but not severe until the summer of 1984. Second, the current episode was by far the most extensive and severe; it extended west of the Great Lakes and was reported extensively for the first time in New Brunswick (308). Third, the dieback continued to progress steadily in Quebec through 1987 (135) but was relatively stable in Ontario (310). As of 1989, there were clear signs of maple recovery in both Quebec (137) and Ontario (148).

This brief historical overview indicated that region-wide, progressive forest crown diebacks were highly episodic (Fig. 1):

i) Onset years were abrupt and region wide; the progression of crown dieback within the region varied in speed and severity that related to local site influences.

ii) Each major dieback episode included a "lead" species and 5-30 associated tree species. Recent episodes typically included tree species affected in earlier major diebacks.

iii) Recovery from crown dieback was equally abrupt and dependent on local factors.

iv) Once initiated, at least a chronic or background level of dieback persisted on the lead and most associated dieback species. Since 1950, dieback has occurred in every year on one or more tree species in some locales of the northern hardwood region.

v) Successive major dieback episodes this century have become more severe in terms of number of species, areal extent and (possibly) wood volume losses to mortality.

Causal Mechanism

Meteorological Event Analysis

Temperature and snow conditions during the winter of 1981 were highly variable. Air temperature at Lennoxville, Quebec indicated that severe cold from mid-December 1980 through late January, 1981, was followed by balmy conditions in February (Fig. 2). For 25 days (February 15 to March 2, March 5-March 13), maximum air temperature exceed 0 C; on ten days maxima were greater than 5 C and had risen to 17.5 C on February 23. Minima exceeded freezing on ten days in February. This spring-like condition resulted in sap flow followed by premature bud-break on many trees. Typically bud break is accompanied by sap ascent, rehydration of tissues, and the partial loss of frost resistance (Fig. 2B). As a consequence, the sudden cold snaps on March 3 and March 14-17 likely resulted in severe freezing damage and tissue death.

An analysis of this 1980-81 winter event suggested five likely mechanisms capable of causing extensive tissue damage and mortality. i) The thresholds of frost resistance were exceeded in late December and early January for all tissues. ii) Roots are the least frost resistant of any tree tissue. Sakai and Larcher (389) noted the root tips freeze at temperatures as high as -1 C to -3 C, permanent roots freeze between -5 C and -20 C, and only rarely are freezing thresholds at lower temperatures. An examination of snow depth indicated that 20 cm or more was on the ground during the intense cold of mid-December to late January (Fig. 2C). This was sufficient to prevent soil frost at the instrumented site. However, in forested stands, surface irregularities and, on crests, blowing snow were likely to have resulted in open snowless patches and localized frost penetration deep into the soil. iii) Frost resistance limits were exceeded for newly rehydrated tissues after the February thaw. The intense and prolonged thaw in February was accompanied by sap ascent, rehydration, and partial dehardening of tissues. The freezing events in early and mid-March were among the coldest on record. Freeze injury on forest trees in late winter after growth has begun can be especially damaging to tissues including the cambial and conductive systems (387, 482).

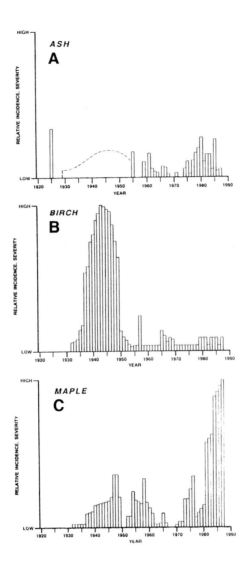

Fig. 1(A) The relative incidence/severity of crown dieback on black, white and red ash, (B) yellow and white birch, and (C) sugar maple hardwoods of eastern Canada reconstructed over the 1920-1987 period from survey reports and published literature.

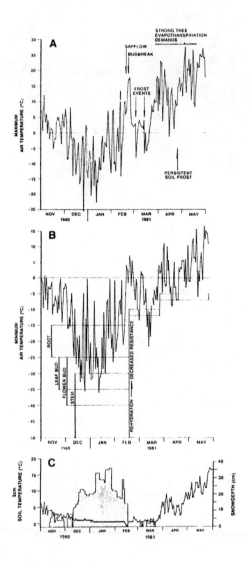

Fig. 2(**A**) Maximum and (**B**) minimum daily temperatures Nov. 1, 1980 to May 31, 1981 at Lennoxville, Quebec (28). Approximate dates of observed sap flow and bud-break are indicated. Representative frost resistance limits are for tissues of temperate deciduous trees acclimatized to regions with severe winters (271). (**C**) Snow depth and soil temperature at 5 cm depth on the Lennoxville meteorological site (28).

iv) The rapid, deep freezing on March 3 and March 14 likely resulted in cavitation of the xylem; in trees where sap ascent has already occurred, sudden freezing of the sap can result in the formation of air bubbles in sapwood as dissolved gases come out of solution (441, 522). v) The complete meltdown of the snow cover over the second half of February and most of March coincided with intervals of low air temperatures (minimum -21.5 C on March 17) and the strong likelihood of deep soil frost penetration. This possibility is reinforced by the fact that the thaw resulted in saturation of the soil profile from snow melt. Wet soils have as much as 50 times the thermal conductivity as dry soils of the same texture. As late as the last week in April, sugar bush owners had noted that the forest soil remained frozen to the extent that farm tractors could be driven over hard ground (D. Lachance, Personal communication 1990). Unusually warm conditions in late March, April, and early May at a time of persistent soil frost would have damaged tree tissues by acute frost desiccation and by preventing the normal reversal of cavitation. This reversal depends on a resurgence of sap by positive root pressure. The advent of root mortality would have both minimized any possibility of active root pressure and contributed to impaired water uptake later in the season as soils thawed (441).

Were the onset years of the birch and ash dieback also typified by the "false spring" pattern observed in 1981? Although the specific winter and spring daily temperature records show 1925, 1937, and 1981 to be highly individual, they had in common at least one pronounced thaw-freeze event. All had 2-4 weeks of intense cold in the winter, and exceptional episodes of early warm weather followed by severe frost in the late winter or early spring.

The magnitude of the winter thaws in 1925, 1937, and 1981 was unique. In each case the stream discharge of the Chaudiére River watershed in the area of severe dieback exceeded 325% of the monthly normals (Fig. 3). We inferred the thaws in these three years were sufficiently prolonged and intense to have resulted in greatly increased risk of frost and cavitation injuries leading to dieback.

Xylem Injury by Cavitation

Chronic injury to the xylem is likely to have resulted both from the sudden freezing of sap as the return to winter-like conditions ended a prolonged February thaw and from acute frost desiccation in late April and May. Both conditions can result in severe cavitation of the vessels and tracheids in sapwood (470). Sperry et al. (441) recently demonstrated that the presence of cavitated sapwood in sugar maple growing in Vermont was closely related to weather conditions; 89% of small branches and 69% of the trunk had cavitated during

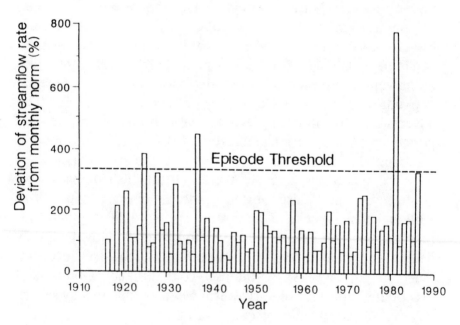

Fig. 3. Winter stream hydrology on the Chaudière River watershed uniquely identified the onset years of major forest dieback episodes in northern hardwoods. Onset years (1925, 1937, 1981) had pronounced winter (D, J, F) thaw resulting in snow melt, and stream flow in excess of 325 % of the normal monthly rate.

cold, sunny conditions in February and 31% of the main trunk had cavitated as a result of only moderate water stress by the end of August. Normally, cavitation is reversed as air bubbles are dissolved or forced out of branches under positive sap pressures created by roots (and in <u>Acer</u> species, by stem tissue) in the spring. They observed this to occur quickly after a rain in mid-March. They also noted, however, that some sugar maple trees had not reversed cavitation as late as May and June and suggested that frozen soil and roots may have been a factor preventing the normal resurgence of sap. Trees in this condition undergo massive and perhaps irreversible cavitation. Irreversibility may result also from the injury or mortality of the ray parenchyma by frost, desiccation, or infection (513), extensive root kill by frost, and/or the formation of tyloses in some species as the roots locked in soil ice fail to replace water loss over a prolonged period of high evapotranspiration demand in spring and summer. In cavitation-injured trees, water reaches the canopy only with difficulty and the trees become hypersensitive to drought and heat stress. The dry, hot conditions in the summers of 1982 and 1983, for example, were accompanied by rapid development of crown dieback symptoms on sugar maple in southern Quebec.

Is there evidence of irreversible cavitation damage in trees with crown dieback? There were no actual on-site measurements of cavitation on any of the severely affected tree species in the 1925, 1937, or 1981 dieback episodes.

The only exception was a few selected experiments on the effect of "vapor blocks" on water transport in declining yellow birch (153). The exception, however, proved to be a crucial piece of evidence supporting our perceptions on the foremost importance of an extreme climate-cavitation-dieback mechanism. This link was the most specific and robust of the causal mechanisms of forest dieback yet proposed. It was remarkably consistent with and broadly supported by the factual evidence, foremost of which was the following: i) Massive cavitation of the trunk and branches was observed on dying yellow birch with severe crown dieback (153); an updated interpretation of the "vapor blocks" and anomalous patterns of water-dye movements in the sapwood confirmed that the symptoms were definitely those of a xylem cavitation injury (470). Swanson (452) noted white birch was ten times more sensitive to cavitation than an associated species, trembling aspen (Populus tremuloides Michx.) when subject to acute frost desiccation in the laboratory. ii) Pathology investigations into the cause of jarrah (Eucalyptus marginata Donn ex Sm.) dieback in western Australia indicated that the only anatomical anomaly was the presence of more root xylem vessels with tyloses in dying trees than in unaffected trees (487). A review of the sporadic jarrah diebacks of 1921, 1928, the late 1940s, 1950s, 1964, 1972, and 1982 indicated they were associated with exceptionally heavy rainfall and more specifically with waterlogging followed by drought (107). Recent glasshouse experiments indicated that if jarrah seedlings are waterlogged, the xylem vessels in the tap root embolise, and though quickly sealed off with tyloses, continue to transpire (i.e., cavitate) (108). iii) Recent studies in central Europe (85, 304, 373, 374) and the Pacific Northwest (32, 149, 375) have confirmed the importance of rapid, extreme climate fluctuations such as winter freezing and/or spring frost in inciting crown dieback in conifers. Conifers show a wide range of sensitivities to cavitation under increasing water stress (470). This is contrary to theoretical arguments (522) supported by a few experiments in the mid-1950s under specific laboratory conditions on one or two conifer species. The fact that Abies species are among the most sensitive to cavitation dysfunction (470) was consistent with the frequent crown dieback symptoms observed on the genus in Europe (100), North America (369, 483), and Japan (261). iv) Pomerleau (363) successfully induced crown dieback symptoms on sugar maple and yellow birch in a set of experimental field plots in which deep soil frost penetration was induced by removing any accumulated snow over the winter months. Acute frost desiccation was likely since the soil remained frozen well into late spring at a time of normal canopy development and high evapo-transpiration demand. His series of laboratory experiments on white birch seedlings confirmed both a sensitivity of roots to acute frost desiccation and extreme moisture stress as a mechanism of crown dieback on this species.

Temporal Correspondence

A marked relationship was evident between climate warming in the Northern Hemisphere and forest dieback (Fig. 4). The onset of widespread crown dieback in 1925, 1937, and 1981 on tree species of northern hardwoods in eastern Canada related closely to episodes of rapid increase in northern hemispheric (and global) temperature. The consistency of the global climate change - forest dieback link was illustrated by the following: (i) Each onset year of a major dieback in the mid-high northern latitudes coincided with a period (3-9 years) of rapid hemispheric warming; each successive dieback episode occurred at the point the mean annual northern hemispheric temperature exceeded the previous record mean annual temperature, and involved a new lead species (Fig. 4); (ii) Recovery from symptoms of dieback on affected trees and/or low levels of new dieback episodes were apparent in years below the "dieback threshold" (equated here to the mean annual hemispheric temperature of 1925, the onset year of the first major region-wide dieback); (iii) Severe dieback and mortality on balsam fir (Abies balsamea (L.) Mill (369) occurred in 1954 within the northern-most limits of northern hardwoods. The mean annual hemispheric temperature of 1954 exceeded that of the previous record year (Fig. 4). This conifer dieback both coincided with an increase in dieback symptoms observed on hardwood species in the region, and showed many dieback characteristics of hardwood episodes, namely a marked thaw-freeze in February of 1954.

The link between global or hemispheric warming and an increased incidence of winters of variable temperatures remains untested. It is, however, supported by Wigley's (510) theory that the extremes (i.e. climatic variability) are highly sensitive to changes in the mean. With global warming, there is a tendency for winter (and spring) temperatures to increase more than summer and fall temperatures, and for variability to be highest during warm, dry intervals. The equivalent climatic condition in the tropics and subtropics was a period of heavy rainfall followed by drought. A literature review of the Pacific Rim and Southeast Asian declines (66, 225, 330, 359, 504) indicated crown diebacks were associated with prolonged periods of wet, cloudy weather followed by clear, bright, hot weather systems. It is possible (but as yet unproven) these episodes were within intervals of rapid hemispheric warming. The likely initiating dieback mechanism was rapid growth accompanied by the deposition of relatively thin-walled vessels and tracheids which cavitate under the subsequent interval of high moisture stress and/or, in some species, the cavitation and formation of tyloses in root xylem under soil waterlogging. We noted in several cases, stands with dieback were in soils with water periodically suspended over a hardpan or other impervious layer. The effect was anomalous growth (thin-walled conductive cells, root tyloses) while the soil moisture persisted but sudden water stress as the soils dried out.

THEORY AND HYPOTHESIS

On the basis of the evidence we presented above, we proposed the following theory and hypothesis (Fig. 5):

Theory: Episodic, region-wide forest dieback may be driven by global climate change.

Hypothesis: The universal mechanism among regional and historic episodes of forest dieback is a xylem cavitation injury initiated by extreme freezing and/or moisture fluctuations at the regional and local levels.

Corollary 1: Progression of crown dieback is directly proportional to the degree of (i) irreversible cavitation injury and (ii) of water stress in the growing seasons following cavitation injury.

Corollary 2: Recovery from crown dieback can occur in the absence of contribution stresses (e.g. during years with normal

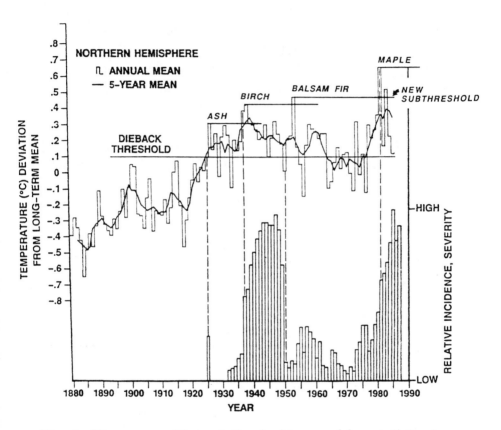

Fig. 4. The compositive relative incidence of forest dieback on ash, birch, balsam fir and maple of eastern Canada in relation to annual and 5-year mean temperatures in the northern hemisphere (28, 165). Dashed lines identify onset years of ash, birch, balsam fir and maple dieback episodes and recovery on birch (1950). The dieback threshold is the mean annual hemispheric temperature of the first episode on ash (1925). Subthresholds are the record mean annual hemispheric temperatures within each dieback episode.

winters and cool, wet summers). The recovery mechanism is the annual deposition of new xylem.

TEST OF UNIVERSALITY

Are the winter thaw-freeze fluctuations observed in the onset of diebacks in northern hardwoods also characteristic of widespread persistent dieback in other regions? Is it possible, for example, that the onset of "pole blight" dieback on western white pine (291) in the Pacific Northwest (1936) and the "new forest decline" on Norway spruce in central Europe in 1979-1980 (236) were triggered by a common phenomenon?

The pattern of winter and spring temperature variability in 1936 at Creston, British Columbia (Fig. 6A) was remarkably similar to that observed during the 1981 event in southern Quebec (32).

Winter thaw-freeze fluctuations in 1978-79 were especially marked in the onset of "Waldsterben" or the "new" forest decline on Norway spruce in central Europe (Fig. 6B). Of the five thaw-freezes between November 1978 and February 1979, two involved temperature shifts of 25-30 C within 24 hour

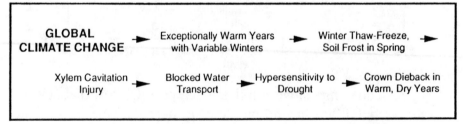

Fig. 5. Conceptual model linking global climatic change to forest dieback in the north temperate and boreal zones. Although forest dieback can occur in the absence of global climate change, the latter markedly increases the probability of extreme climate fluctuations that incite region-wide cavitation injury and crown dieback.

periods and minima temperatures below -10 C. A blocking high in late December 1978 at the onset of the "new" forest decline was the most pronounced on record since systematic 500 mb upper tropospheric pressure records were initiated in 1946 (258). A sudden shift westward in the blocking system brought in exceptionally cold air from the continental interior. We surmise that the freezing, stem cavitation, root frost damage, and mortality typical of the northern hardwood diebacks were very likely. The existence of deep soil frost in spring has yet to be verified but even cold soils at a time of above normal air temperatures in April and May may have contributed to cavitation and/or desiccation injury (389, 441).

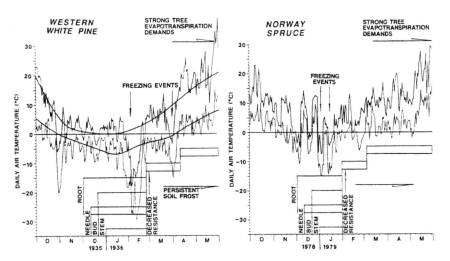

Fig. 6A. An event analysis of the onset of "pole blight" on western white pine (Pinus monticola) in British Columbia and the Pacific Northwest United States. Meteorological records are from Creston, British Columbia.

Fig. 6B. An event analysis of crown dieback in central Europe based on meteorological records at Lahr, West Germany. The event has been implicated in the dieback on Norway spruce (373, 374).

SUMMARY

In this paper, we have proposed a biophysical theory and hypothesis of forest dieback. Our review of the historical literature indicated this precept was central at the very inception of intensive experimental studies on birch in the late 1940s and 1950s (93, 153, 175, 363). It was highly instructive to us that early researchers viewed diebacks as a "physiogenic" rather than a biotic phenomenon and to note that the discussion of probable influences did not involve air pollution but linked forest dieback in some way (as then not fully understood) to a change in climate. Our case study approach complemented and added to these early observations and provided a continuity of conceptual and historical perspective.

A focus of much of the research by atmospheric and forest scientists in the 1980s was on air pollution as a likely cause of forest decline. Manion (292), in critiquing the six pollution-related hypotheses of Schutt and Cowling (410) was convinced that none came close to adequately explaining "Waldsterben" or other forest diebacks. To account for forest diebacks in areas relatively free of regional air pollution (e.g. Pacific Rim), distinctions were made (incorrectly (32)) characterizing those diebacks as symptomatically and causally different from those in central Europe and eastern North America (314). The adoption of a multiple-factor hypothesis to explain the absence of an obvious link between air pollution and dieback ultimately proved to be a perceptional miasma. It was simply too difficult to test because of the high level of complexity and interaction proposed.

The case study approach provided us with a wealth of experimental and observational detail necessary both to substantiate a single-agent mechanism and to develop an encompassing theory on the ecology of forest dieback. Our approach was consistent with three essential steps defined by Manion (292), namely, (i) to search for a simple and single-agent explanation <u>first</u>, (ii) to identify the mechanism of interaction between agent and dieback, and (iii) to structure a narrowly focused testable hypothesis of cause together with a much broader conceptual framework on forest dieback as an injury or disease phenomenon. It is noteworthy that Sinclair et al. (431) recently modified Manion's (291) concept to include the possibility that sudden, single-agent events can initiate forest dieback and "decline".

Our premise that diebacks are associated with xylem cavitation injury was consistent with the ecological evidence from a wide spectrum of forest regions. As a dysfunction of the conductive tissues involved with the transport of water and nutrients to the tree canopy, it was also consistent with the evidence from remote sensing (strong infra-red signal), tree nutrition (leaf element deficiencies or imbalances), and insect (stem borers, leaf defoliators) and disease (Armillaria and possibly Phytopthora) etiology.

Evidence essential to support our theory of a climate-cavitation-dieback link was limited. Foremost was the lack of field measurements of irreversible cavitation injury on trees with severe crown dieback. Second, quantitative estimates of the severity and extent of crown dieback were lacking over the historical period. There was an abundance of descriptive data on crown condition since 1950 but this needed to be systematized and related to climatic records. A test of the central premise that climatic change or related phenomena (128, 338) at the global level increases the probability of extreme climatic fluctuations (at the regional and local scale) would be a significant addition to our theory. Currently the evidence is lacking.

We are developing and refining a biophysical model to estimate the potential risk of crown dieback on selected tree species. This model will be driven by climate and soil hydrologic variables that affect leaf water potential, and ultimately, the degree of cavitation. The intent is to use the model to verify the existence and geographic distribution of past dieback episodes, and to predict, under scenarios of future climate change, what will be the probable incidence and severity of forest dieback.

Disclaimer

Although the information in this document has been funded wholly or in part by the United States Environmental Protection Agency through EPA Contract No. 68-D9-0052, it has not been subjected to Agency review and therefore does not necessarily reflect the views of the Agency and no official endorsement should be inferred.

THE GERMAN FOREST DECLINE SITUATION: A COMPLEX DISEASE OR A COMPLEX OF DISEASES

Otto Kandler

Botanical Institute University of Munich
Menzinger Strasse 67, D-8000 Munich 19, Germany

THE "WALDSTERBEN" CONCEPT DEFINITION, EARLY PREDICTIONS, AND HYPOTHESES

In the early 1980s, various authors claimed that a large scale forest decline was occurring in central Europe, especially in Germany (67, 407, 408, 410, 412, 471, 472, 473). These claims, quickly popularized by the media under the slogan "Waldsterben" (forest death), maintained that:

i) All the important tree species were involved and showing similar symptoms, not recognized before.

ii) The syndromes stand apart from ordinary diseases of trees.

iii) The main symptoms were decrease in increment, loss of foliage and feeder roots, discoloration (yellowing) of foliage, premature senescence, altered branching habit, and abnormally high seed production.

iv) Waldsterben developed very rapidly. For instance, Norway spruce (Picea abies (L) Karst.) and pines showing first symptoms were expected to deteriorate rapidly enough to require cutting within 3 years (410) and Norway spruce and silver fir (Abies alba Mill.) in the Black Forest were expected to die off within 10 years (403).

v) The most likely cause of Waldsterben was an ecosystemic complex disease triggered by cumulative stress from increasing air pollution and deposition (e.g. acid rain, sulfur dioxide, nitrogen oxide, ozone, hydrocarbons, etc.) (408).

FOREST HEALTH SURVEYS IN GERMANY

In West Germany, since 1983, the extent of the alleged "novel forest damage" leading to Waldsterben has been estimated by annual forest surveys employing a classification of the forest condition into four very generally described damage classes based on foliage deficit (crown transparency) and foliage discoloration (Table 1). The number of trees in the various damage classes is converted to the corresponding forest area they cover. Thus the annual damage reports of the German authorities give the percentage of the total forest area covered by trees of each damage class. Using the series of seven annual forest surveys (1983 to 1989), it is possible to judge the credibility of the Waldsterben scenarios and the reliability of the prognoses on development, published in the early 1980s.

Table 1. Classification scale of the annual forest damage surveys in Germany

Damage Class	Loss of Foliage in %
0 = healthy	< 10
1 = slightly damaged	10 to 25
2 = distinctly damaged	25 to 60
3 = heavily damaged	> 60
4 = dead	

Early prognoses assumed a continuous increase in damage allegedly brought about by dose-response relationships for air pollutants. They were based on the survey results of 1983 and 1984 which indicated a steep increase of forest damage (Fig. 1). However, the 1983 survey was not performed by the same methods in all states of West Germany (248), and the field personnel were not fully acquainted with the new classification scale.

In contrast to the predicted fast progress of damage and death, the percentage of the various damage classes of the forest area of West Germany leveled off and remained almost constant since 1984 when uniform survey methods were fully established. There has been no progression from lower to higher damage classes, as expected for a syndrome leading toward death (Fig. 1).

The absence of the predicted continuous rise in classes 3 and 4 has recently been attributed to "salvage" or "stress oriented" harvesting (99). However, there has been no significant extra cut resulting in an excess of available timber or a sustained fall in timber prices as predicted by the Waldsterben scenarios (342). Also, the total annual tree harvest corresponds to only 1 to 1.5% of the total forest area and, as such, could not prevent a rise in damage class 3 and 4 if the predicted progression in damage did occur.

The cumulative presentation of forest damage of all tree species in the total forest area of West Germany depicted in Fig. 1 disguises not only the annual fluctuations of crown conditions of single trees, but also the species and site-specific differences in the development of damage classes. However, such information is epidemiologically important and is needed as a basis for rational hypotheses on the etiology of the damage.

The annual fluctuation of crown conditions was best studied in Switzerland where the degree of foliage deficit of ca. 7500 marked trees, included in the annual surveys (14), were judged in stepped 5% damage classes. The percentage of trees of each 5% class exhibiting an annual change of >10% has been calculated (Fig. 2). The majority of the trees show improvement, in the higher damage classes 2 and 3. Twenty to thirty percent show no change, and only 10-15% deteriorate further. Since only 1.3% of the total number of trees belong to damage class 3, annual mortality is below 0.5% and thus lies close to the normal range of annual mortality of 0.1-0.4% in managed forests (394). This value is distinctly below the annual mortality of 1-3% in unmanaged forests (10), which is extremely rare in Europe but common in North America.

Damage to species varies from state to state and over time as depicted for damage classes 2-4 in the most southern (Bavaria) and the most northern (Schleswig-Holstein) states of West-Germany (Fig. 3). In Bavaria, the order of the species based on the degree of damage was pine > Norway spruce >

beech > oak in 1983/84, but it changed to oak > beech > Norway spruce > pine in 1989. Similarly, inversions of the order of species and dramatic non-synchronized fluctuations in the extent of damage are found in Schleswig-Holstein (Fig. 3B) and the other states (not shown) and also in much smaller areas within states (234). Such behavior is not compatible with a general air pollution hypothesis, since damage to the tree species should reflect the different species-specific sensitivities to prevailing pollutants or mixtures thereof, and responses should remain the same unless the concentrations or kind of pollutant change significantly.

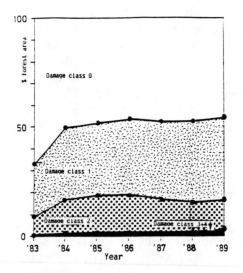

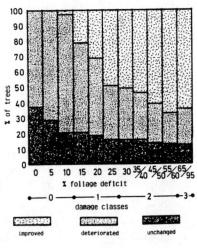

Fig. 1. Percentage of forest area damage classes 1-4 for 1983-1989 for all tree species in the Federal Republic of Germany.

Fig. 2. Percentage of ca. 7,500 trees from Swiss annual forest surveys in 5% foliage deficit classes remaining unchanged, improved >10%, or deteriorated >10% during the period 1987 to 1989. (14) The relationship of foliage deficit to FRG damage classes is provided at the bottom.

However, no correlations exist between the fluctuations of air pollutant levels in rural areas of West Germany and that of damage in the various tree species. The pH of rain water showed a slight decrease in the late 1960s but increased again to the original level in the late 1970s (Fig. 4), the beginning of alleged Waldsterben. Nitrogen oxide and ozone levels remained virtually the same over the period (Fig. 5).

High SO_2 concentrations have caused a widespread dying of conifers in industrial areas and city centers in the first half of this century (cf. 237). However, thanks to effective measures taken in the 1960s, SO_2 concentrations have decreased to levels below 50 $\mu g/m^3$ even in the most industrialized Ruhr area, while SO_2 concentrations in rural areas remained rather constant at levels below 15 $\mu g/m^3$ (Fig. 6). As a consequence, the Ruhr area and city centers have become recolonized by lichens and conifers (235, 509). At present, the only large-scale SO_2 damage in forests in central Europe is found in the Ore Mountains and neighboring mountain ranges at the German-Czechoslovakian border. It is caused by the SO_2 emission of a series of power plants immediately south of the Ore Mountains. The correlation between the erection of the power plants and the dying of spruce trees on the crest of the Ore Mountains has clearly been demonstrated by increment measurements in stands in different distances from the first plant which was started in 1953 (479).

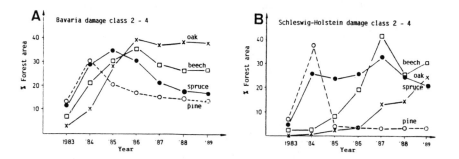

Fig. 3. Percentage of forest area in damage classes 2-4 for four tree species in the most southern (Bavaria) and northern (Schleswig-Holstein) states of the Federal Republic of Germany for 1983-1989.

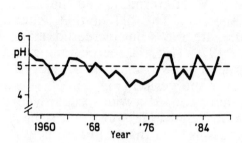

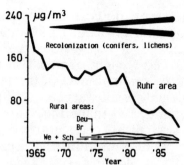

Fig. 4. Longest record of pH in rain water in a forest area of central Europe (Monitoring station Retz, Austria; (90).

Fig. 6. Annual means of SO_2 concentrations in highly industrialized (Ruhr area) and in rural areas (stations) as in Fig. 5) of Germany (19).

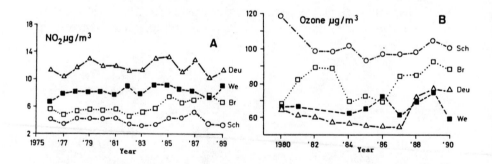

Fig. 5. Annual means of NO_2(A) and ozone (B) measured at four permanent monitoring stations in rural areas of the Federal Republic of Germany. Brotjacklriegel (Br), 1016 m asl = Bavarian Forest; Schauinsland (Sch), 1205 m asl = Black Forest; Deuselbach (Deu), 480 m asl = Rheinland-Pfalz; Westerland (We), 4 m asl = Schleswig-Holstein. Data from Umweltbundesamt (Berlin).

CROWN TRANSPARENCY IS NOT A NEW SYMPTOM, NEITHER QUALITATIVELY NOR QUANTITATIVELY

The observed tree-, species-, and site-specific fluctuations of crown conditions are typical phenomena of the normal differentiation processes caused by inter- and intra-species competition within the stand and by spatial and regional variations of climate and other site conditions. These are combined with at least partly reversible species-specific, often endemic, diseases of biotic origin. Unfortunately, the recent annual forest surveys cannot be compared to corresponding data from former periods, because the present classification system of forest damage was not used before 1983. This dilemma is clearly expressed in Landmann's (270) final comment at the International Congress on Forest Decline Research in Friedrichshafen, West-Germany, October, 1989:
"Saying that levels of defoliation are "high" or suggesting that the observed evolution is not normal, implies a reference to such "normal". Even if one has good reasons to think so, the only valid conclusion is that it is impossible to state whether the levels of defoliation and current fluctuations are normal."

Actually, there were numerous reports about poor forest conditions in central Europe early in this century (336). However, attempts to quantify the degree of crown transparency over a large area are rare. One of the few relevant estimates of forest damage in a large region is that of Rebel (371) who observed an increase in very poor crown conditions in conifers and hardwoods in Bavaria early in the 1920s and called this phenomenon "heat disease". He assumed that a series of drought years had caused this conditions. According to his judgement, 21-51% of the forest area of Bavaria was "endangered" (damage class 1?) or "irreversibly diseased" (damage class 2-4?), depending on the growth area. In the same period, Wiedemann (507) reported an annual loss of timber production of 20% in the Norway spruce forests of Saxony, and a wide-spread "Fichtensterben" (Norway spruce death) with unknown causes was reported in East Prussia by Hitschold (201) ten years later.

Many other reports on regional and local diebacks, often comprising all tree species and thus resembling the present description of Waldsterben, are found in the forest literature through the century. A most vivid example is the description of a dieback on a granite ridge in the Black Forest by Hiss

(200): "Here we have not only "Tannensterben" (fir death) but also "Buchensterben" (beech death). A closer look shows also "Fichtensterben" (Norway spruce death). ... The heart of the forest is sick. Its vitality has reached a level where dying begins...". Today this particular site shows predominately 20 to 60 year-old stands in normal condition. However, a few old fir trees still exhibit short pieces of broken limbs hidden behind many adventitious branches thus indicating that trees have suffered in former decades.

Although the symptom description is vague in the old reports, recent photographs documenting Waldsterben (e.g. 48, 407, 412) are similar to those of trees and stands found in old scientific journals, photograph albums and postcards. For instance, of about 90 photographs published by Rubner (383, 384), almost none of the spruce trees presented as typical of particular "mountain races" would today appear in class 0 (healthy) but would be put in classes 1-3. Even trees depicted on old realistic paintings of the 16th century show crown conditions resembling the so-called novel forest damage (101). Comparison of old and new photographs of the same stands or groups of trees show the fluctuation of crown conditions but do not indicate a recent general decline (234, 521).

Photographs also provide a quantitative evaluation of the frequency of crown transparency in former times for comparison with current conditions. Four experts classified crown conditions for Schweingruber (413) according to the presently used methods of the annual forest surveys of about 2000 clearly recognizable Norway spruce trees on old (before 1927) and recent (1975-1987) postcards collected from the Swiss Alps and other areas of Switzerland. The estimates for trees in damage classes 2-4 were 16-41% and 16-21% on the pre 1927 and 1975-1987 postcards, respectively, as compared to 16-18% for the 1984 to 1986 annual Swiss forest surveys.

Another source for objective information on the extent of former crown thinning are the studies in the first half of this century by Burger (81) on leaf area indices (LAI) of 167 normal (at least to then applied standards), 40 to 150 year-old Norway spruce trees from 35 stands at different locations in Switzerland. Burger's data, recently converted into crown transparency by Bucher (79), show a wide variation (Fig. 7). When the trees are classified into four transparency classes ($0 = < 0.35$; $1 = 0.36-0.5$; $2 = 0.51-0.8$; $3 = > 0.81$), the proportions of trees by the defined transparency classes 0 to

3 are 52%, 28% 17% and 3% respectively. If Bucher's classes 0 to 3 are roughly comparable to the classes 0 to 3 of the current West German forest survey, then the current forests have about the same proportion of healthy and damaged trees (Fig. 1) as compared to the proportion of the earlier Swiss sample.

Burger's data also show a distinct correlation between crown transparency and the elevation at which the tree was located. A similar correlation is shown by the damage classes of the recent annual surveys and is considered as an indication of photooxidative damage, since photooxidants increase with elevation (21, 365). The alleged increase in tree damage associated with increases in air pollutants is therefore not substantiated by the comparison of the earlier sample population with current surveys, even though crown transparency in both populations is correlated with elevational increases in photooxidant air pollutants.

INCREMENT MEASUREMENTS DO NOT INDICATE A RECENT GENERAL GROWTH DECREASE BUT RATHER GROWTH IMPROVEMENT

Annual ring analysis is the most objective tool for retrospective studies on tree vitality. Thus, annual ring analysis of a few pairs of healthy and damaged Norway spruce were compared in the very beginning of the Waldsterben discussion. A distinct decrease in annual increment was found in the damaged trees of most of the comparisons (132). However, in the following years, when not single pairs but representative collections of trees were studied, only a very loose correlation between annual diameter increment and damage classes has been observed. The majority of the trees showed no recent growth decrease but normal or even optimal growth compared with earlier in the century.

The most extensive example of more recent increment analysis is the study of 900 Norway spruce trees as part of the 1987 Swiss Sanasilva report (14) (Fig. 8). The growth reduction caused by drought years in the mid 1970s was followed by quick recovery. By the early 1980s, the majority of the trees were growing at rates of up to 20% above the index line, and even the 8% of the trees in damage class 3 had nearly returned to index level growth.

Current average or above average tree diameter increment

and timber production per hectare also was reported in the joint Switzerland, Germany and France evaluation (53). When the growth curves of different generations of Norway spruce stands are compared (Fig. 9), the 30 and 60 year-old stands were found to be growing faster than the 90 and 120 year-old stands when they were the same age (53, 398). The improved growth in the 1980s is incompatible with the Waldsterben concept which was based on the assumption of a general decrease in increment, a hasty extrapolation from an increment decrease observed during the drought years of the mid 1970s. Now the opposite question is gaining momentum; why do forests grow faster in the second half of this century than they did in the first?

Nitrogen input by deposition of air pollutants (e.g. nitrate, ammonia), elevated carbon dioxide, and slight increases in precipitation and temperature are frequently assumed to contribute to improved growth (53). However, improved forest management such as regular thinning, melioration before planting, etc. may be more significant growth promoting factors. In former times, removal of litter and humus for agricultural usage contributed to reduction of timber production by 10 to 30% (370). Significant increases in growth should have occurred following discontinuation of this poor practice in the middle of this century.

Although most forest stands have normal or improved growth, some have poor growth and substantial death. These are mainly restricted to orographically extreme sites with unfavorable climatic conditions and poor mineral supply. These site conditions along with endemic diseases and insufficient forest management lead eventually to decline and diebacks as had been observed in earlier times, but the exact causes of such "natural diebacks" (328) are not fully understood as yet.

WALDSTERBEN: A PROBLEM OF AWARENESS

Comparison of old and new photographs, the frequency of crown-thinning calculated from old leaf area index data, growth ring chronologies, and the results of annual surveys in the period from 1983 to 1990 all suggest the same conclusion: General dying of forests (Waldsterben) or even an unprecedented decline in central European forests during the 1980s has not occurred. Therefore, there is no reason to

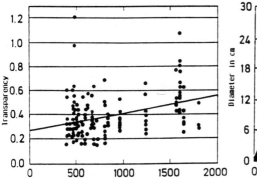
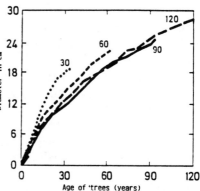

Fig. 7. Distribution of crown transparency among 167 Norway spruce trees depending on elevation (m asl). The data have been calculated by Bucher (79) on the basis of leaf indices determined in Switzerland before 1950 by Burger (81).

Fig. 9. Increase in stem diameter of 30, 60, 90 and 120 year-old populations of Norway spruce in stands with similar site conditions. n = 47 (398).

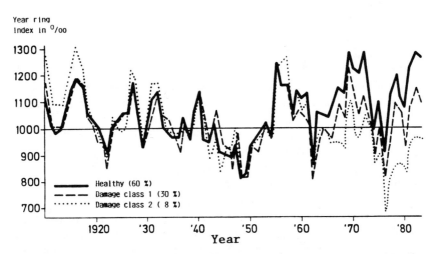

Fig. 8. Year ring chronology of 900 Norway spruce trees also included in the Swiss annual forest damage surveys. Years ring index corrected for age trend (Sanasilva 1987); (14).

suggest a new pollution-triggered complex disease, and thus, Waldsterben may be understood as a problem of awareness. Forest conditions that were believed to be "normal" in earlier times became suddenly a symbol of the growing fear of the destructive potential of human activities on the environment. Patterns of suboptimal crown conditions and oscillations in incremental growth, associated with variations in site, weather, diseases, and infestations, have clearly been observed in the past, but as they continue to occur in the 1980s, they have been incorrectly recorded as novel forest damage.

A main component of the complex of diseases of the central European forests are the well known root and heart rot diseases caused by various fungi and combinations thereof which lead eventually to crown thinning and/or discoloration (421, 432) described as the principal symptoms of Waldsterben. No attempt has been made to exclude these endemic diseases from the totals of the annual surveys even though root diseases occur widely and may reach levels of 100% infection in older stands (395). Only once, in the 1989 survey of the province of Bozen (Northern Italy), was stem rot monitored separately. Although a rather simple method was used and only severe cases could be recognized, a clear correlation between damage classes and stem rot was found. As shown in Fig. 10 almost 30% of the trees classified in damage classes 2-4 had stem rot, while only 6% of the damage class 0 (healthy) trees had stem rot. Examples of the fluctuation in crown conditions in Norway spruce associated with infections by <u>Heterobasidion annosum</u> and <u>Armillaria</u> spp. have been documented by photographs taken at different times (234, 236). These findings indicate that stem rot is a significant but seldom properly noted component of the novel forest damage reported in the annual surveys.

Among the many other less common diseases influencing crown conditions, two symptomatically well defined, but with respect to etiology not fully understood diseases, should be discussed in more detail. These syndromes have been highly popularized by the media into the symbols of Waldsterben. One of the syndromes is a specific form of decline of silver fir, commonly called "Tannensterben", and the other is a special type of yellowing of Norway spruce associated with magnesium deficiency. The symptoms, development, and history of the two syndromes are so different that is obvious that they represent two independent diseases rather than two variants of one complex disease.

TANNENSTERBEN, A WELL DOCUMENTED SPECIES-SPECIFIC SYNDROME

Tannensterben, recognized in the 19th century, was first documented by photographs (Fig. 11) early in this century (337). It is characterized by increment decrease, needle loss, dying of branches from the bottom to the top of the crown, and the formation of a "stork's nest" on the top. A long lasting decrease in increment, preceding the first well documented episode early in this century, started in northeastern Bavaria in the 1880s (508). Up to 80% of the silver firs were diseased in the early 1920s. Most of the diseased trees died or were cut. Thus, silver fir is rare in these areas today.

The spreading and episodical outbreaks of Tannensterben has been vividly described by Claus (94) (our translation): "A few decades ago, the fir stands in the districts of Tharandt -(he names 12 districts of Saxony) - still appeared in good condition. Today only odd specimens of firs exist in these districts, and most of these trees carry the mark of death in their crowns. First reports on massive fir death in Saxony came from Judeich and Manteufel in the year 1875. Death was so rapid that Judeich, even at this early date, raised the question of whether it was still worthwhile to continue planting fir. ...Later reports, around the year 1886, from Bohemia, Silesia, Bavaria and Wurttemberg, support the assumption that Tannensterben extended concentrically from Saxony. At present (in 1928) we have complaints about Tannensterben from almost all countries of central Europe."

Tannensterben became so popular in the 1930s that the concern for the health of silver fir entered the phraseology of the Nazi party to strengthen nationalistic feelings as reflected in the following quotation of Graser (152) (our translation): "Forests have always had an important influence on our national history and spiritual heritage, an influence which is especially deeply rooted in the inner life of our people. It is therefore the duty of every generation of foresters to preserve and hand on intact to their descendants and thus to a permanent organic Germanic-German culture this basis of the German emotional experience with forests and trees in all its various aspects. For this the health of our silver fir is essential."

In view of the well documented history of Tannensterben and its former popularization, it is difficult to understand why this old syndrome is now considered to be a component of the novel forest damage. Although the most recent report of the German expert committee on Forest Damage and Air Pollution now admits that Tannensterben is not a novel syndrome (21), it is still claimed that it had never before exhibited such a wide distribution and intensity, and thus is suggested to be a new phenomenon caused by a complex of factors, with anthropogenic air pollution playing the key role. This is an unusually careless treatment of a well documented spreading syndrome which was considered an epidemic of unknown causes by former authors (c f. 126, 337, 508). The epidemic character of the most recent outbreak has been correctly characterized by Malek (290) who writes (our translation): "With regard to its geographical distribution, Tannensterben appears as a series of explosions, which burst periodically and simultaneously within the whole central European distribution area, at its margin as well as in those regions generally considered optimal for fir. The outbreak of dying recurs periodically."

The most recent episode of Tannensterben began in the 1960s. It became widespread in the Black Forest and in Switzerland (53, 414) where it has been reported only as a local disease at the beginning of this century. It was also widespread in the Bavarian Forest (260) where Tannensterben had not been reported before, although heavy outbreaks of Tannensterben had been previously reported in other parts of Bavaria (94, 508).

Increment decrease started in most cases between 1960 and 1970 (Fig. 12). Minimum increment, combined with increased mortality, occurred in the middle of the 1970s when the syndrome was reinforced by a series of drought years (238). Recovery of incremental growth started in the early 1980s, and normal annual increment was reached again in the late 1980s. Changes in crown conditions were delayed several years. Crown transparency reached its maximum in the mid 1980s, while recovery by the formation of adventitious branches occurred in the second half of the decade. A decrease in the percentage of damage classes 2-4 from its maximum of 82% in 1985 to only 58% in 1988 was reported in the Bavarian annual forest damage report (249).

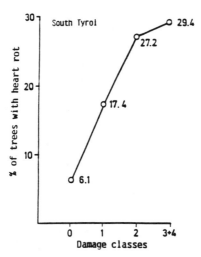

Fig. 10. Relationship between heart rot and damage class in Norway spruce (Southern Alps; Annual forest survey 1989, province Bozen).

Fig. 11. First photographic documentation of Tannensterben" (silver fir decline) in the Fichtel Mountains Bavaria. (*) = Single Norway spruce trees which would be classified in damage class 2 according to present standards (photo: Neger, 1908,(337).

The increase in ring width occurred simultaneously along the stem from the bottom to the top (Fig 13A) and started at a time when crown conditions were still very poor (Fig. 13B). The formation of adventitious branches lagged behind the onset of renewed cell division in the cambium, thus suggesting that it is unlikely that recovery of increment resulted from increased photosynthates produced by an enlarged leaf area. Instead, one may assume that enhanced cell division and the respective change in allocation of photosynthates may have been triggered by a change in hormonal supply. For instance, the formation of cytokinin in root tips, essential for cell division also in the stem (cf. 299), might be severely repressed in the diseased tree and restored after the disease is overcome.

It is noteworthy that not all fir trees in a stand became diseased. In both districts of the Bavarian Forest, 15 to 20% of the trees investigated showed no significant increment decrease (Fig. 12) but merely the usual annual variations in increment typically associated with climatic conditions. This observation and frequent recovery are compatible with a disease hypothesis but do not support the pollution and climate hypotheses, although climate certainly exerts modifying and synchronizing effects on disease development.

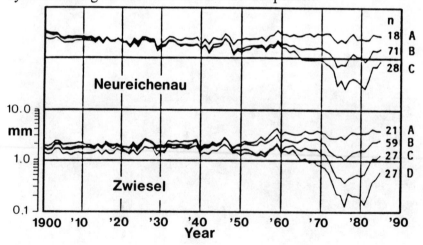

Fig. 12. Year ring chronologies of silver fir from two forest districts in the Bavarian Forest. Cores were taken at random from dominating and condominating trees. The samples of each district have been grouped according to the depth of the annual growth reduction (A-D; n = number of trees per group) (237).

Attempts to recover root pathogens have usually been made from trees exhibiting heavily damaged crowns. Negative results obtained by many authors, beginning with Neger (337), may be due to the delayed expression of crown symptoms. Recent experience (236, 238) indicate that trees with heavily damaged crowns may have already overcome the disease as indicated by the increase in annual ring width (Fig. 13). Thus, further attempts to elucidate the causative agent of Tannensterben should concentrate on trees in the early stages of increment depression rather trees with advanced crown damage.

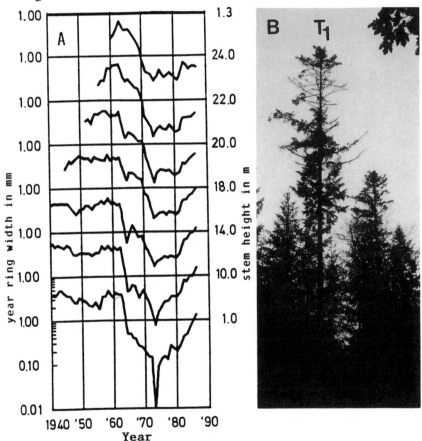

Fig. 13. Year ring chronologies in different heights of a silver fir (T_1) exhibiting extreme symptoms of "Tannensterben" (A). Heavily diseased silver fir (T_1) shortly before it was cut and analyzed (B). (237).

"ACUTE YELLOWING" OF NORWAY SPRUCE - A NOVEL TYPE OF Mg DEFICIENCY

In contrast with crown transparency and other symptoms of Waldsterben, "acute yellowing" of Norway spruce is considered by most authors to be a new syndrome ("novel forest damage"), observed only since the early 1970s. This condition has been variously termed "needle yellowing in higher elevations of the German Mittelgebirge" (21) and "mountainous yellowing" (265).

Acute yellowing differs from other types of yellowing by being limited to older needles, beginning at the tip and progressing to the needle base. Yellowing preferentially occurs on the upper or illuminated side of needles. The syndrome is restricted to mountain areas with acid soil derived from granitic, sandstone and other rocks poor in available Mg.

Acute yellowing accounts for less than 5% of the total Norway spruce damage in Germany, but because of its eye-catching appearance, rapid expansion in the early 1980s, and popularization by television reporters, acute yellowing together with Tannensterben are considered by the public as the most spectacular expressions of Waldsterben.

The condition has been diagnosed as a Mg deficiency by demonstrating close correlations among needle yellowing, Mg content of the needles, and available Mg in the soil (71, 518, 527) and a fast regreening upon application of Mg containing fertilizers (524).

Two hypotheses for acute yellowing based on assumed increases in air pollution have been developed: i) The ozone/leaching hypothesis assumes that ozone and/or other photooxidants trigger oxidative damage to cell membranes and chloroplast pigments. These in combination with acid rain enhance leaching of minerals. Because of the low Mg level in the soil, the plant cannot compensate for the loss of Mg (365, 366). ii) The acidification/Al toxicity hypothesis assumes an acidification of the soil by deposition of pollutants. This would lead to an enhanced cation leaching and a shift of the Ca/Al ratio favoring Al and subsequent inhibition of Ca and Mg uptake by the roots (472, 473). Both hypothesis still lack direct experimental proof under ecologically relevant conditions. In addition, fumigation experiments were unable to demonstrate lowering of Mg content of needles by the combined action of ozone and acidified fog (265).

The development of acute yellowing in forests during the past decade is not consistent with either of the pollution based hypotheses. Increased amounts of yellowing would be expected, based on dose-effect relationships of continuous exposure to relatively constant annual levels of air pollutants (Fig. 4, 5, 6), but acute yellowing shows an episodic development with frequent spontaneous regreening and a low rate of mortality (239, 240, 241).

As an example, Fig. 14 shows a section of a natural regeneration Norway spruce plot SP II from the Bavarian forest where yellowing started in 1984. In 1985, 10 additional trees yellowed while 13 of the original 94 yellowed trees regreened. Two yellowed trees in 1984 had no additional yellowing in 1985 (stagnating yellowing). Regreening dominated in subsequent years. By 1989 less than one percent of the 1600 marked trees of plot SP II were still diseased.

The development was similar in plot SP I, 1 km from plot SP II, where yellowing had started 3 years earlier. Yellowing was much more severe with about 50% of the trees diseased in 1985. By 1989, most of the trees had regreened and resumed normal growth rates except for a few scattered single trees.

However, yellowing continued on 40% of the trees in a 100 square meter section of the plot SP I even into 1990. Such "nests" of yellowed trees were observed earlier in the decade in older stands in other mountain areas of Germany (233). They often appeared as centers from which yellowing expanded. In older stands, trees in such "nests" today have significant needle loss, frequent root and heart rot, and higher mortality (240, 241). These may represent situations where acute yellowing is superimposed on other diseases or may be reinforced by especially poor site conditions.

The dynamics of acute yellowing, characterized by non-synchronous yellowing and regreening of single trees within the same natural regeneration was corroborated by the Bavarian forest survey reports 1986-1988 (Table 2) (249). The Bavarian annual surveys are based on more than 70,000 older trees at 1430 observation plots. Of the trees classified in yellowing group 2 and 3 in 1985, 70% and 54% respectively were classified as healthy in 1988. On the other hand, 33% of the trees classified in yellowing groups 2 and 3 in 1988 were classified as healthy in 1985. Thus, during the period 1985 to 1988 approximately 700 trees recovered while only 108 trees became yellow. Only 38 trees out of 988 trees of yellowing classes 2 and 3 in 1985 had died or had been felled within the period 1985-1988.

Since yellowing is correlated with Mg deficiency in the needles and is restricted to areas with soils of low available Mg, one would expect that spontaneous regreening is correlated with increased Mg content of the needles and increased Mg content of the soil. Increased Mg in regreened needles has been reported (239, 241), but there are no data relating the episodic course of acute yellowing and respective changes of available Mg in the soil. On the contrary, soil samples from the rooting area of healthy and yellowed trees of the same stand showed no significant differences in pH or ammonia exchangeable Mg (239, 241). Thus, low levels of available Mg in the soil appears to be a site-specific predisposing rather than a tree-specific inciting factor associated with the wave of acute yellowing observed on Mg poor soils in the 1980s. Furthermore, none of the young hardwood trees, shrubs or herbs growing in the mapped natural regeneration plots SP I and SP II showed any symptoms of Mg deficiency, thus indicating that the inciting factors of acute yellowing is species-specific.

Considering the episodic nature of acute yellowing, its rapid geographical expansion, and the small-scale mosaic of healthy, diseased and spontaneously regreening trees, acute yellowing resembles an epidemic disease. Uptake of minerals from the soil might be limited either by an antagonistic component of the soil microflora at the level of the rhizosphere or mycorrhizae or by root diseases. Thus, some trees would show characteristic symptoms of Mg deficiency on sites where Mg availability nears deficiency for healthy trees. This view is supported by recent experiments where Norway spruce were grown in soil from healthy and diseased stands from different areas of central Europe. Various treatments with steam, fungicides, and inoculation with mycorrhizal fungi suggest the presence of a noxious microbial factor in the soil from diseased stands that interferes with the rhizosphere and/or ectomycorrhizae thereby inhibiting mineral uptake (112, 120).

Therefore, acute yellowing is most likely a syndrome of complex etiology. In extreme cases, it may be caused by too little available Mg in the soil or by severely restricted Mg uptake caused by root diseases or antagonists of rhizosphere flora or mycorrhizae. In most cases, low Mg availability in the soil is the predisposing factor and the antagonistic microbial component is the inciting factor. Unfortunately, such phytopathological interactions between the soil microflora and mineral availability or uptake have rarely been discussed (cf. 432). Such interactions may not be restricted to the particular case of acute yellowing of Norway spruce but may be an important factor in other cases of fluctuating mineral deficiencies.

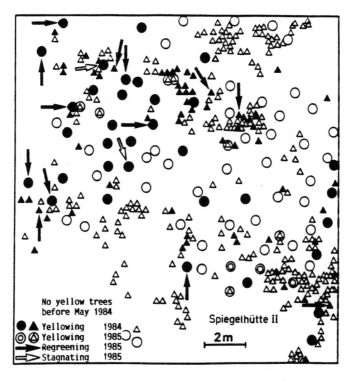

Fig. 14. Section of a mapped natural regeneration of Norway spruce at Spiegelhütte (Bavarian Forest) showing the population dynamics of "acute yellowing" during the first two years of yellowing. All trees were regreened in 1988. (241).

Table 2 - Changes of the yellowing classes of > 60-year old labelled trees in the 1430 observation plots on which the annual surveys in Bavaria are based. (249)

In 1985	in yellowing (class 2)	912 trees
In 1988	recovered (class 0)	70.0%
In 1988	improved (class 1)	13.0%
In 1988	worse (class 3)	1.4%
In 1988	dead or felled	2.8%
In 1985	yellowing (class 3)	76 trees
In 1988	recovered (class 0)	54.0%
In 1988	improved (class 1 or 2)	18.5%
In 1988	unchanged (class 3)	9.2%
In 1988	dead or felled	18.0%
In 1988	yellowing (class 2 + 3)	329 trees
In 1985	not yellow (class 0)	33.0%
In 1985	class 1 (progressing to 2 or 3)	34.0%
In 1985	already in class 3 (unchanged)	4.6%

LONG LASTING GROWTH REDUCTIONS AND DECLINES
A POORLY UNDERSTOOD COMPLEX OF UNSETTLED TREE DISEASES

Decline syndromes of unknown causes, such as the above described Tannensterben, discoloration and loss of foliage, crown dieback and other decline symptoms (295) are common in all tree species. At present, oak and beech decline attract more attention in Germany than Tannensterben, because damage in these two species has increased over a period of time when conifers are generally characterized as recovering (Fig. 3 A, B). Earlier decline episodes in these hardwood species are frequently described in the forestry literature. Unfavorable climatic and soil conditions, poor forest management practices, pathogens, insects and combinations of these are suggested as causative factors for these declines (oak: cf. 111, 114, 335, 346, beech: 385).

Air pollution has rarely been discussed as a causative factor for hardwood decline, since the rather high resistance of hardwoods to the pollutants is readily evident. Hardwoods showed growth inhibition to some extent but have survived in industrial conurbations and city centers. Only recently have air pollutants been assumed to play a role as a predisposing factor, mainly via soil acidification, in the decline of oak (174) and beech (380). However, no direct evidence justifying this assumption is known, and with the exception of classical smoke damage, no correlation has been demonstrated between the spatial and temporal distribution of any known air pollutant and hardwood decline.

Suggested causes for the various unsettled tree declines of Germany are vague. Most interpretations assume that these problems are complicated ecosystemic complex diseases with emphasis on environmental stress as a primary factor (473) and biotic agents as secondary factors (214, 408). However, during the past 150 years, forest pathologists have successfully elucidated the etiology of a number of diseases that were originally termed "declines". Specific biological inciting agents have been found for Dutch elm disease, chestnut blight and a host of problems. Why should we assume that there will be no further progress in identifying the causal pathogen of some of the still unsettled decline diseases? Is it appropriate to accept Houston's (214) statement that the existence of unsettled decline syndromes "testify to the fact that for them no causal pathogens, capable of inciting diseases in healthy trees exist"? Such a conclusion resembles the reasoning of the Waldsterben advocates when they suggest that air pollution has a role in "novel forest damage" because "no other hypothesis likely to explain convincingly the wide extend of damage in several species when pollution is not involved could be provided" (11). Such proofs based on negative evidence have never led plant pathologists to a true understanding of the etiology of any disease.

A warning example, not to accept pure environmental arguments, is provided in the history of the elucidation of declines of vegetatively propagated long-lived plants like grafted fruit trees, grapevines, poplars, and potatoes. These declines were originally described as degenerative processes (150) but became more popular under the term "Abbau" or "running out" (323). Although mechanical transmission of symptoms in the 1920s and electron microscopic observations of potato virus X were made in 1936, biotic agents were considered to be of only secondary importance until the late 1940s by many authors in Germany and Russia. Complex environmental conditions and anthropogenic influences were thought to be the primary causative factors. The prevailing and most popularized hypotheses were based on terms and suggestions such as: i) "Ecological decline" is due to continuous vegetative propagation under unfavorable site conditions (323). ii) "Productivity stress" is caused by modern cultivation techniques and fertilization (254). iii) "Deviation of the metabolism" is caused by cultivation under climatic conditions different from the plant's native country (69). iv) "Redox shift in the soil" favors oxidizing conditions thus causing potato decline (376). v) "Ecological phase-shift" favors premature aging, according to Lysenko's developmental-phase doctrine (cf. 315).

None of these ecological hypotheses, which contain some elements of the present hypotheses for the so-called Waldsterben, has proved valid. Today, phenomena formerly described as decline ("Abbau") in agricultural crops and fruit trees have been recognized as a complex of diseases caused by biotic agents. Instead of fungi, the "classical" pathogens of plants, viruses, mycoplasma-like organisms (MLO) and in some cases Rickettsia (rather unconventional pathogens for forest pathologists) have been shown to be inciting factors in these previously mysterious decline diseases. For instance, in potato decline, more than twenty viruses, three MLOs, and one viroid are involved. They often occur in various combinations in the same plant and exhibit synergistic or at least overlapping effects. The resulting complex syndromes confused phytopathologists for decades and gave rise to the cornucopia of ecological hypotheses mentioned above.

The role of these unconventional pathogens in forest health is poorly understood. However, MLO, first discovered in mulberry only 20 years ago, have been shown to cause epidemics in deciduous fruit and forest trees. These are accompanied by decline phenomena (cf. 199, 296). The number of known MLO diseases is still increasing as shown by the recent discovery of MLO-associated diseases in about 10 deciduous trees in central Europe (cf. 416). Among them was alder (<u>Alnus glutinosa</u> (L.) Gaertn.) decline (417) which had been suggested to be a component of novel forest damage (408). Therefore, MLOs must now be considered as potentially common pathogens of hardwoods in central European forests. Further progress on the understanding of the epidemiology and etiology of MLO diseases is expected once more sensitive techniques, such as DNA-probes (262), are available to detect MLOs in tissues with greater precision.

Although viruses have frequently been found in deciduous trees and more rarely in conifers (cf. 89, 341), only a few viruses produce leaf symptoms in forest trees. However, virus diseases are common in fruit trees causing not only leaf symptoms but also decline phenomena. Even latent virus infections may reduce growth and productivity. Highly managed forests initiated with nursery growth stock are particularly vulnerable to virus initiated declines because of the ideal conditions for spread and accumulation of viruses in the nursery, especially when trees are propagated by cuttings or tillers or via tissue cultures. Therefore, intensified virological research in forest trees, up to the level common in fruit trees, probably would disclose involvement of virus infection in decline diseases.

CONCLUSION

The central dogma of the Waldsterben concept suggests a quasi synchronous occurrence of unprecedented decline in all tree species in central European forests since the 1970s due to a complex disease of forest ecosystems triggered by air pollution. However, a decade of research on novel forest damage in West Germany has shown that: i) The symptoms considered to be specific for the new complex disease, such as needle loss in conifers, acute yellowing in Norway spruce, loss and discoloration of foliage in hardwood species, have not evolved concurrently. ii) Periods of increasing damage and recovery alternate independently in different species and regions. iii) Year ring chronologies show no unusual inroad in increment in the 1980s. iv) No spatial and temporal correlation between novel forest damage and air pollution could be shown. v) Retrospective studies on forest condition suggest similar levels of crown transparency in Norway spruce at the beginning of this century as today and recurrent decline episodes in the main tree species. Thus, the results of a decade of research are not compatible with the assumption of a new ecosystemic complex disease triggered by air pollution. They rather confirm the occurrence of non-synchronous fluctuations of forest conditions and recurrent episodes of clarified as well as unsettled species-specific declines. These are described in the literature and considered to be caused by a complex of diseases, infestations and abiotic stresses. This aspect is emphasized by Landmann's (270) statement at the end of the International Congress of Forest Decline Research at Friedrichshafen; "To sum up, the consideration of all the observed problems as a single syndrome stems from a perception or a conviction but is not a scientific conclusion".

ACKNOWLEDGEMENTS

I am indebted to Dr. P. Manion for helpful suggestions and critical reading of the manuscript.

Data from the author's laboratory discussed in this paper are taken from the project "Epidemiology and Etiology of Fir and Spruce Decline" commissioned to the author and financed by the Bayerisches Staatsministerium für Landesentwicklung und Umweltfragen.

A CLOSER LOOK AT FOREST DECLINE:
A NEED FOR MORE ACCURATE DIAGNOSTICS

John M. Skelly

Department of Plant Pathology
The Pennsylvania State University
University Park, PA 16802

During the 1980's, more attention has been given to the status of forest health than ever before. This immense concern should lead to more prosperous forests for future generations. Much of this recent interest has been fostered due to several dieback and decline situations in European and North American forests which have been perceived as being unprecedented. Nowhere was the concern more intense than in the Black Forest (Schwarzwald) of southern Germany where both silver fir (Abies alba, Miller) and Norway spruce (Picea abies (L.) Karst.) displayed rather dramatic symptoms of crown thinning and needle yellowing. In eastern North America similar reports of "unprecedented" declines emerged concerning high elevation spruce-fir (Picea rubens Sarg., and Abies balsamea (L.) Mill. or Fraser fir (A. Fraseri (Pursh) Poir.) forests (75, 263) and sugar maple (Acer saccharum, Marsh.) (313).

Reports of these "unprecedented" declines have since resulted in a massive amount of popular and scientific literature. Considerable lay person and scientist concern has been created by reports that presume regional and worldwide synchrony of occurrence for declines and that ascribe, a priori, involvement of anthropogenic air pollutants in these problems. Numerous factors must be considered before such broad and all-too-important conclusions are reached. As forest biologists, and indeed as air pollution specialists, we must constantly be mindful of "normal" forest pathological, entomological, and abiotic stress occurrences in our forests. Excellence in the scientific method must be maintained. Scrutiny of all other pathogenic agents prior to concluding an etiological role for air pollutants and/or global scale environmental changes inclusive of complex interactive studies must be completed before concluding a predispositional role for air pollutants in exacerbating any specific decline etiology.

A NEED FOR ACCURATE DIAGNOSIS OF CAUSAL AGENTS

Within human clinical medicine, the diagnostic procedure conducted by the doctor is of utmost importance to the eventual well-being of the patient. Simple diseases are regularly diagnosed and prescriptions for their control or symptom alleviations are offered. More complex disorders may be referred to specialists for more detailed diagnosis and subsequent treatment procedures. In most instances, with the exception of the very young or otherwise communication impaired patients, the medical doctor queries the patient for additional important information before reaching a diagnostic conclusion.

Plants (forest trees) cannot offer verbal assistance in the diagnostic efforts to determine the etiology of their respective maladies. Hence, the plant (forest) pathologists are left to their own skills, knowledge, and techniques to ascribe causes of disease based for the most part upon visible symptoms and/or signs of biotic pathogens.

Many commonly occurring diseases and insect pests have been described in forests (e.g. 173, 188, 231, 431, 437). Maladies incited by biotic pathogens and insects are occasionally easy to ascribe as to cause due to classical symptoms and sometimes presence of the actual organism on or within the affected tissues. Cultural techniques may be employed to further assist in proper diagnosis of biotic plant pathogens. Never the less, it is important to recognize that symptoms in forest trees are not indicative of etiology. Numerous symptoms in forest trees are unclear. The ephemeral nature of grazing types of insects and the paucity of fruiting structures of causal organisms make diagnosis in the field quite difficult. Abiotic agent induced diseases are sometimes more difficult to assign to a single agent due to the commonality of symptom expressions induced by differing causal agents. In the case of diseases caused by multiple agents, such as may be involved when forest species exhibit symptoms of dieback and decline, the diagnostic procedures likewise become more specialized and difficult.

An exception to these disease prognoses has occurred recently in the manner within which forest responses to anthropogenic air pollutants have been supposedly diagnosed. For the most part, proper diagnostic procedures have been ignored in lieu of claims of extraordinary forest declines and dieback being due to acid rain and/or its air pollutant precursors. As offered by Chamberlin (91), "Too often a theory is promptly born and evidence hunted up to fit in afterward. Laudable as the effort at explanation is in its proper place, it is an almost certain source of confusion and error when it runs before a serious inquiry into the phenomenon itself".

Numerous articles continue to offer credence to "unprecedented" and "unexplained" forest decline phenomena via perpetuation of incorrect information within general introductory sections of the respective articles, e.g., (86, 247, 343, 392, 451). Each of these articles justifies studies reported therein by terminologies such as "...increasing concern... about the declining health of forests in central Europe..." (86); "increasing awareness of forest dieback in Europe...", (343); ..."widespread forest decline observed in West Germany," (247); "Forest declines have been widely observed in central Europe..." (451) and ... "it has now been accepted that a new trend in the decline and dieback of forests is becoming increasingly apparent in western Europe and on the North American continent" (392). Bach (33) proceeded even further by stating that his paper addressed the generally accepted premise that forest dieback is a complex phenomenon caused by multiple stresses that are exerted by a host of contributing factors. Without citing etiological or diagnostic proof of air pollutant involvement, Bach (33) also calls for a pressing need to introduce active control measures for air pollutant emissions. These statements have been made in the respective introductions of the cited articles without scientific references to the purported diebacks and decline actually taking place. In addition, each of these articles, in an introductory sense, suggests that acid rain and/or its pollutant precursors has been demonstrated to play a role in the "observed" forest decline. These articles, as well as numerous others, then proceed with details of the respective study in a manner very acceptable for publication. However, the idea that a widespread forest decline exists in central European and North American forests has become perpetuated. Furthermore, any defined connection with anthropogenic air pollutants on a regional scale (e.g., acid rain) must be viewed with considerable scrutiny. This criticism was made by Ballach and Brandt (38) when they reviewed the BML (Bundesminister für Ernährung, Landwirtschaft und Forsten) forest decline inventory for 1983. They noted analysis of causes without the proper differential diagnostic procedures. Air pollutants were the only reported causes of injury, yet much damage was obviously also due to a widespread drought in the summer of 1982.

Throughout the 1980s (8, 9) and into the 1990s (189), one of the most popular scenes depicting acid rain and/or air pollutant-induced damages within our environment involved photographs of dying and/or dead trees. Acid rain has not as yet been identified as directly inducing any type of symptom on any forest tree species under ambient conditions (24), and even ozone has not been shown to directly kill forest trees as casually illustrated by Hewitt et al. (189). The science community should thoroughly evaluate and review the forest health diagnostic procedures employed by the authors before accepting such overwhelmingly biased photographic evidence of mortal effects.

The need for diagnostic accuracy in determining causes of any observed change(s) in forest and tree health has never been more important. If we are to accurately judge changes in forest or tree health as due to anthropogenic air pollutants or global scale alteration of the environment, we must first be capable and mindful to account for naturally-induced perturbations. We must also be able to discern the numerous ways in which all factors (including anthropogenic influences) interact with one another in leading to each distinctive phenomenon such as individual tree species or an entire forest "decline" situation (293, 435, 436).

FOREST DECLINE TERMINOLOGY

Decline and dieback are terms used to describe a

pathological symptom complex involving growth reductions, leaf size or number losses, and twig and branch necrosis which sometimes leads to death of the entire organism. Manion (291) and Houston (213) simply describe forest tree decline as a gradual and general deterioration leading to death. Causes may entail the interactions of a number of abiotic and biotic entities which cause stress within the individual tree over some indefinite period of time. Symptoms of decline may be subtle but progressive. Periods of decline have been followed by recovery which may be either temporary or complete depending upon the spatial and temporal scenario of implicated causal agents.

The term "forest decline" should be considered a misnomer. Even though it has been frequently used along with forest dieback, or mortality, a more correct term may be "forest species decline". Similar terms of "Waldsterben" (forest dying), "Waldschäden" (forest damage), or "neuartige Waldschäden" (new-type forest damage) as they exist in German literature are likewise incorrect. The term "Waldsterben" was popularized by Schütt and Cowling (410) as a collective term for a group of simultaneously occurring symptoms on several major forest tree species in central Europe. Within their oft-cited article, Waldsterben has been offered as a new specific disease, as if some cataclysmic phenomenon is taking place across distant forests. Since evidence of diagnostic proof was not offered, one could easily also conclude that many of the disorders of the trees pictured within their article had common, well-known, and easily detected causes. Nutrient deficiencies, spruce spider mite (Oligonychus ununguis, Jacobi) and even rime ice are among the most easily noted potential etiological agents. Yet they summarized "Never before have so many tree species, growing under so many different soil, site, and climatic conditions shown so markedly similar and serious effects."

Entire "forests" have seldom been demonstrated to decline, rather perhaps a predominant species within the "forest" may have suffered dieback, decline, and mortality. Millers et al. (319) present an overview of numerous declines and diebacks of eastern North American species which have occurred over these past 100 years. Many of these forest tree species suffered similar symptoms but numerous known and different causes were identified. In these instances of "forest" decline several species may have been noted to decline but entire forests on a regional basis were not affected.

"Forest" decline should be replaced by the more correct term, "species" decline. Declines of forest tree species have been reported to occur over large portions of a species geographical range (319, 331); many of these declines have documented causes. An even more accurate description of "species" decline would sometimes follow as site related and species specific decline. Examples of these types of declines may include decline and mortality of red oak (Quercus rubra L.) and scarlet oak (Q. coccinea, Muenchh.) on dry shaley soils (442, 445), littleleaf disease of southern pines on poorly drained soils with clay hardpans (517) and recently reported crown yellowing and general decline of Norway spruce on sites replete of Mg and/or other required nutrients (239, 244, 525). Diagnosticians of these site specific species declines should have a much more detailed description of site variables which may influence approaches taken during respective diagnostic procedures. Individual trees should be carefully examined for evidence of signs and symptom relationships, isolations of would-be biotic pathogens, insect activities, and site requirements of nutrition, water, soil characteristics, and similar parameters.

Several recent publications (243, 410, 411, 474, 514) serve as examples of nondiscriminant use of decline-type symptoms across many different forest tree species as lumped into the term "forest" decline or "Waldsterben". Numerous pictures of symptomatic foliage and dying and dead trees appear in these publications with minimal listing of their potential causes. Yet, implications of causality, if by nothing more than their appearance in these air pollutant effects oriented articles, is inferred for the reader in an a priori fashion.

Subsequently, articles more popularly written for the general public have included more and more dramatic presentations of devastated forest scenes (253, 286, 316, 364). Air pollutants directly, or under the "synonym" of acid rain, have been linked in each one of these publications as major incitants of the forest damages which have been pictorially displayed (Fig. 1) (263). Furthermore, the term "Waldsterbensymptomen" has been coined as a new catch-all description of any form of unhealthy tree condition. Fig. 2 clearly shows a Norway spruce under obvious stress -- but from what causes? The authors (411) propose these symptoms, from Burlington, VT, to be similar to Waldsterben appearances as evident on this same species in central Europe. More careful diagnosis would have to consider: spider mite infestation, cytospora canker (Cytospora spp.), eastern spruce gall adelgid (Adelges abietis, L.), heat injuries from a nearby roof, mechanical damages to stem or roots, chemical injuries from salt or herbicides and/or drought and winter freezing injuries. Only complete examination would reveal such etiological agents.

Symptoms as observed on silver fir in the early 1980s, have been shown in Fig. 3 (263). Numerous agents including drought and mineral nutrient deficiencies, as well as increasing air pollution, have been suggested as a cause. Numerous references attest to long-term nutrient depletion (primarily magnesium and potassium supplies) in the Black Forest (124, 525). Concurrent studies of Mg deficiency of Norway spruce have been reported in northeastern United States by Ke and Skelly (244). Most severe increases in damages were recorded

Fig. 1. Cover photograph of widely distributed publication depicting scenes of forest devastation. (Krahl-Urban et al., 263). (Karl Peters, Photographer).

107 *Picea abies* mit Waldsterben-Symptomen
n Burlington, Vermont, USA

Fig. 2. Figure 107 from Schütt et al. (411) illustrative of non-specific symptoms of Norway spruce but referenced as "Waldsterben-symptomen".

on silver fir and Norway spruce during the period 1983-1986 within the Black Forest following the dry years of 1980, 1982, and 1983 (365, 405). Recovery of damaged trees of the early 1980s has developed since 1985 (523). Recovery was photographed in late fall of 1989 (Skelly, personal observation) (Fig. 4).

Kandler (236, 238) offers a far differing opinion of the causes of the species and site specific symptom expressions being seen in German and central European forests. He offers pictorial evidence of symptoms present in the early 1900's on both silver fir and Norway spruce throughout the regional forests and further concludes that many recent reports of concern demonstrate a lack of awareness of past occurrences of similar symptoms and epidemiological considerations.

Severely damaged, middle-aged silver fir in Munstertal at an elevation of 900 m. Its growth is very reduced, and the tree is weakened by mistletoe. This tree will probably die in the next couple of years.

Fig. 3. Photo from page 91 Krahl-Urban et al. (263) showing silver fir near Freiburg, im. Br., FRG. This tree was expected to have died, "in the next couple of years". Photo taken in 1983 (Karl Peters, Photographer).

Symptoms of sub-top dying and crown thinning were noted in forest records in the late 1800s for these two species. Kandler (238) further states that symptoms have not occurred in a synchronous manner across numerous European forest species as suggested by Schütt and Cowling (410). Fluctuations in the major and complex syndrome of crown transparency appeared to closely reflect climatic conditions, especially drought years, and then on a very site specific basis.

Similar examples of a lack of awareness and/or the potential importance of other causal agents and their respective role in decline initiation exists in North American forest situations. In all of the abundant literature dealing with the recent but not unprecedented, (e.g. 13) decline of red

Fig. 4. Evidence of recovery in silver fir during the period 1986-1989. Note new crown emerging along main stem and at the very top of the two outside trees. Photograph is of the same area as depicted in Figure 3; the two trees on the right may be the same tree but photographed from a different angle. (By author, November 1989).

spruce and Fraser fir and balsam fir, few papers cite the significant presence of the balsam wooly adelgid (<u>Adelges picea</u>, Ratz.) throughout the decline areas. Bruck (75) reported the presence of the adelgid on an average of 41% of the Fraser fir across 35 plots in the Black Mountains of North Carolina in 1985. A later survey of 1988 (presented in the same paper) reported an increase to an average of 69% of the trees infested at this same elevation. Studies by Zedaker et al. (519) and Zedaker and Nicholas (520) further identified the importance of the balsam wooly adelgid within high elevation spruce-fir forests in two of three geographic regions they surveyed. Decline patterns were coincident with recent weather phenomena (339) and the occurrence of the adelgid; fir mortality was highly correlated with balsam wooly adelgid infestations. Similarly, Dull et al. (115) concluded that patterns of mortality of the spruce-fir forest of the southern Appalachian Mountains were consistent with previous and continuing infestations of the adelgid. Mortality of red spruce was not considered as alarming with the highest mortality at 14% on Mount Mitchell in the Black Mountains. An average of 49% mortality of fir was recorded for their studies in all plots at all elevations. Bruck and Robarge (76) also attributed an observed increase in mortality of red spruce which occurred in 1987 to the combined effects of a severe regional drought in 1986 and a severe rime ice event that occurred in the Black Mountains on December 11, 1986. Similarly Nicholas and Zedaker (340) have reported that balsam wooly adelgid infestations have spread to virtually all remaining live firs in the Great Smoky Mountains. Percent infestation has increased from 46% in 1987 to 99% in 1989. Concurrently, they have observed that the crowns of 35% of the red spruce residual in the now open stands had visibly deteriorated between 1985 and 1989.

During this same period, Bruck et al. (77) dismissed their previous reports of intensive adelgid activity at their research sites. They did so with a single statement within the abstract of the 1989 paper. "With the exception of the balsam wooly adelgid, few signs or symptoms of disease or insect attack were noted on either Fraser fir or red spruce populations". The major portion of their paper (77) proceeded with presentation of data and discussion of acidic deposition by cloud water impaction at the high elevation sites. Other papers (e.g. 392) further elaborate upon the chemical climate of the Mount Mitchell area and evaluate its potential for inciting the "forest decline". They state in their introduction ... "the red spruce and Fraser fir forest has registered decline beginning in 1980, the severity of the decline being undisputedly noticeable above frequently observed cloud base at 1585 m MSL." No mention was made of the balsam wooly adelgid and its previously well described devastating effects.

Yet, the most widely distributed photographs of the recent "forest decline" in North America are of the devastated high elevation spruce-fir forests at Mount Mitchell (285) (Fig. 5). Most often these photographs and others of more northern forests have captions implicating, if not directly stating, air pollutants and/or acid rain as the pathogens of greatest importance (243, 285, 286, 480).

The photograph (Fig. 5) which appeared in MacKenzie (285) is actually of the devastation due to balsam wooly adelgid attack of the spruce-fir forest on Mount Mitchell, NC. The caption offers no reference to the actual cause of the

Air pollution is taking a heavy toll on U.S. trees and crops. This heavily polluted spruce-fir forest on Mt. Mitchell, North Carolina, appeared healthy in the early 1980s.

Fig. 5. Photograph and accompanying caption of the spruce-fir forest at Mount Mitchell, NC as offered to the readership of a World Resources Institute publication by MacKenzie (285).

stand decimation but indicts air pollutants as "taking a heavy toll on U.S. trees and crops". During a visit by the author to the Mount Mitchell spruce-fir forest in 1987, the presence of the balsam wooly adelgid was easily discerned (Fig. 6). Furthermore, an overview of the forest showed clear evidence of where the adelgid had been previously controlled up until 1974 (Fig. 7) and where no control had been instituted (Dr. Chris Eager, personal communication, 1987). A brief summary of the adelgid control programs of the 1960s and 1970s in the Mount Mitchell area has been presented by Dull et al. (115).

FOREST HEALTH SURVEYS

During the past decade, considerable attention has been given to monitoring the health condition of forests throughout the temperate regions of the world. Reports from the central

Fig. 6. Stem of Fraser fir exhibiting the presence of the balsam wooly adelgid on Mount Mitchell, NC. White dots show a moderate infestation of the adelgid. (Photo by author).

European forest area as well as from Canada and the United States abound in the literature and have become too numerous to cite. Recent review articles are available which summarize many of their individual and composite findings (39, 73, 80, 222, 423, 483). Additional new reports from Mexico (465) and China (284) indicate a more global interest and concern for forest health appraisals. Such monitoring will, no doubt, continue into the twenty-first century. As a result, the total community interested in forests should have the best-ever information available which offers broad perspectives on forest condition.

Due to its premier location and likewise historical significance, the Black Forest of the State of Baden-

Fig. 7. A view from the Tower atop Mount Mitchell, NC looking SW across the spruce-fir forest. Stand to the left was not sprayed during the 1960s and 70s; stand to the right was protected with insecticide applications repeated on a ca. three-year basis through to 1974. Slow spread via wind eddy's has caused an increasing infestation of the residual live stand to the right. Red spruce are not attacked by the adelgid; but remaining stems are left in the open where the elements of weather and increased insolation take their toll.

Wurttemberg, in southwestern Germany has become one of the most intensively studied and monitored forests in the world. Soon after completion of annual surveys, reports of forest condition are prepared (282) and quickly disseminated to forest personnel as well as being popularized for the awareness of the general public (22, 305). Results of the 1989 forest condition survey for the entire then Federal Republic of Germany were released as part of a scientific conference entitled, "International Congress on Forest Decline Research: State of Knowledge and Perspectives" held 2-6 October 1989 at Fredrichshafen, Germany; a pictorial sketch of forest condition classes appeared soon thereafter in newspaper articles, Fig. 8 (404).

From the information presented in Fig. 8, it can be noted that for 1989 a slight percentage increase in Class 3 and 4 trees (strongly damaged and dead) coupled with a 0.5% decrease in healthy trees (without noticeable damage) has been reported. This would lead to a conclusion that some slight deterioration of forest health had occurred over the 1988 survey results. It is significant that no details on cause(s) of these changes were offered; however, air pollutant induced effects were directly implied.

Fig. 8. Graphic display of forest (tree) condition classes for the Federal Republic of Germany Forest Survey 1983-1989. Reprinted, by permission, from DIE ZEIT/Wolfgang Sischke. Evidence of damage classes is based upon percent defoliation and percent yellowing of four tree species evaluated across the whole of West Germany.

INADEQUACY OF FOREST CONDITION SURVEYS

It is important to understand that European surveys of forest (tree) health have relied upon ground observation with the use of binoculars to estimate foliage losses (percent defoliation) and changes in coloration towards chlorosis (percent yellowing) in mature tree crowns. Plots (trees) have been visited usually only once each year (222). Defoliation has been used as a catch-all term to include leaf drop, presence of abnormally small leaves, fewer leaves produced, and most importantly -- missing parts of leaves. Such crown transparencies may be due to insect, biotic pathogens and/or mechanical damages caused by rime ice, hail, or high winds. Yellowing of foliage as a symptom of altered forest tree health is likewise used as a non-etiologically discriminant value and could be present as a result of drought, premature senescence, nutritional disorders, and numerous biotic pathogens and insect activities.

Fig. 9. Crown of European beech showing 35% crown thinning (defoliation) as observed near Rhinefelden, FRG. (By author, September 1989).

The European beech (<u>Fagus sylvatica</u> L.) illustrated in Fig. 9 was evaluated as being 35% defoliated (Skelly, personal observation) when observed in early September 1989 near Rhinefelden, Germany. However, once a branch was collected (for a separate nonrelated study), in-hand observation revealed several disorders which accounted for the 35% defoliation (Fig. 10). Causes of damage included: beech leaf beetle (<u>Rhynchaenus fagi</u> (L.) evidenced by the round holes; leaf anthracnose (<u>Gnomonia</u> spp.), as evidenced by the marginal and midrib necrosis; adult cicada (<u>Typhlocyba cruenta</u> H.-S.) feeding, as evidenced by the leaf surface scarification; and activity of an unknown Lepidopterous insect, as evidenced by the feeding scars along the margins of leaf (173). Yellowing of leaves was due at least in part to powdery mildew infection of the lower leaf surface. No air pollutant induced symptoms were noted, yet defoliation estimates of 35% for the tree crown appeared valid.

Fig. 10. Close up leaves on branchlet removed from tree in Figure 9 illustrating presence of four injuries obviously leading to the 35% defoliation as recorded for entire crown. (By author, September 1989).

The silver fir trees depicted in Fig. 11-14 are even more indicative of problems encountered in final data interpretation concerning tree health in the forest surveys of the state of Baden-Wurttemberg. These photographs were taken in October 1989 (Skelly, personal observation) of Forest Baden-Wurttemberg (FBM) survey numbered trees within a high elevation silver fir plot above Kappel im. Br., Germany. Large main stem and root collar wounds (Fig. 11 and Tree No. 87 in Fig. 13) were evident. A recently renumbered tree that had suffered a lightning strike three years previously (Tree No. 86, Fig. 13) was still being observed within the survey. The crowns of these trees were reevaluated in the 1989 survey (Fig. 12 and 14) and the data remain as part of the tree condition data presented by the respective damage classes illustrated in Fig. 8. In addition, dead branches evident in 1989 were classified as defoliated in 1989 even though their death obviously took place some 4-6 years previously. Until dead branches fall from these and numerous other survey trees, their previous defoliation occurrences will likewise be a continuing negative influence on the tree condition data presented in popular and even scientific publications.

CONCERNS FOR ON-GOING AND FUTURE SURVEYS

Several activities of forest surveys designed to evaluate forest condition (health) become of importance to the interpretations of annual survey results:

i) Survey crews change due to commonly encountered seasonal employment conditions even though crew leaders or supervisors may be retained on a project for several years.

ii) Training of survey crews in disease and insect evaluations is generally inadequate and involves recognition of only major insect, disease, and abiotic agents which most commonly are found in the general forest habitats. Uniquely occurring agents and the more endemic levels of leaf grazing by insects, for example, are lost into the general figures of percent defoliation and/or yellowing.

iii) In addition to recording only the easily recognized symptoms of foliage and crown disorders, it is important to realize that these diagnoses are made

Fig. 11. Silver fir, high elevation survey plot near Kappel, im. Br., FRG with major wound (canker?) involving nearly 50% of circumference at one meter above the ground. A number 9 (89) shows on upper left side of tree. (By author, November, 1989).

Fig. 12. Upper crown of the tree number 89 (Fig. 8) with numerous dead branches and new shoots on main stem. Note new crown emerging at very top. (By author, November 1989).

    via ground visual estimations or via binoculars where only lower leaf surfaces and the bottom of the crowns are most easily seen. Even then, surface features of damaged leaves are not easily seen nor are they, therefore, accurately evaluated.

iv) No branch samples containing symptomatic foliage are brought to the ground for actual hands-on or hand lens evaluation of potential etiological agents. Diagnostics, meager as they are, are most often carried out without specimen availability.

v) The amount of time allocated per plot is too little to complete a diagnostic-etiological survey.

vi) When present, and even if recorded on field tally sheets, the known causes of leaf loss and/or leaf yellowing are generally not carried forward into the figures released to the forest industry and general public; only the percent defoliation and yellowing are used to place trees into condition classes as illustrated in Fig. 8.

Fig. 13. Silver firs, tree number 86 and 87 of high elevation plot near Kappel, im. FRG, with large basal wound (logging?) evident on tree number 87; tree number 86 had been struck by lightning in 1986 or 1987 as evidenced by cracking and streak up side of tree. Both trees had been recently renumbered for continuation of survey. (By author, November, 1989).

Fig. 14. Crowns of tree (86) and 87 (left) showing numerous dead branches from previous years of stress, but both silver firs have evidence of recovery as new crowns emerge along the upper main stem. (By author, November 1989).

Several of these concerns have already been overcome via the methodologies outlined in the newly emerging forest health monitoring surveys (2, 39).

CONCLUSIONS

The purpose of this paper has not been to underestimate the importance of air pollutants in altering the health of our forest trees, stands, species, or ecosystems. A recent review by Garner et al. (143) attests to important direct and indirect effects of various pollutants on the forests of eastern North America. Similar reviews of European forest conditions have been published which give credence to the importance of understanding pollutant-induced effects (374, 379). Nor has this paper been written with the intention of pointing to the inadequacies of past and/or ongoing forest (tree) health and condition surveys. Rather, the point to reiterate Manion (292) is one of caution in interpretation of survey results.

The use of tree crown defoliation and foliage yellowing data may provide useful information concerning trends of tree health on a yearly or multi-year basis. However, if one is to ascribe cause of such observed symptoms to any single or more complex scenario of agents, one must use far more complex diagnostic procedures over and above ground based observations. These two broad response parameters tell us little as to their incitant and/or of their subsequent effects to the entire organism. Furthermore, simply ascribing the occurrence of these two broad symptoms toward further implications of "forest decline" is biologically irrelevant without evidence of their cause and/or long term effects. These discrepancies become even larger when considering tree to tree, tree within stand, stand within forest and forest within region variations of potential symptoms and their causes.

Recent publications have appeared which indict air pollutants as being the direct cause of "forest" decline (23, 402, 409) or species decline (514). Using circumstantial evidence of symptom synchrony across many species in a temporal sense, lack of evidence of causal agents being present (in surveys only defoliation and yellowing are carried forward in final figure presentations), and spatial co-relationships of pollutant loading and observed damages, these and other authors have concluded air pollutants to be the main cause of "forest decline". However, other authors call for considerable caution when concluding causes of forest (tree) declines (38, 64, 222, 283, 435, 436). Binns and Redfern (64), perhaps, gave the earliest strongly worded admonition, "There is undoubtedly an element of neurosis involved in the readiness of many foresters and some workers to attribute any decline or dieback in forest tree species to the combined effects of atmospheric pollution and acid rain without adequate critical investigation."

As a final point concerning the current status of European forests, we must consider a most recent report (73). Several conclusions offered by these authors include: i) growth studies have indicated a complex relationships between defoliation and increment, ii) there is no evidence of a widespread decline in the growth of European forests; to the contrary, there is limited evidence of a growth increase, and iii) declines as observed in some areas of Europe started much earlier than the late 1970s or 1980s and most appeared to be triggered by extreme climatic conditions. Similar reports indicating no major unexplained or unprecedented declines taking place in the United States have recently been published (39, 423).

Several words of caution must also be heeded so as to not dismiss air pollutants as being of considerable importance to forest health and productivity. Ahrens et al. (1) have recently reported on a growing data base on tropospheric ozone as it occurs throughout remote sections of the Black Forest; potential long-term effects remain undetermined. Also, as pointed out by Garner et al. (143), this well known and phytotoxic air pollutant should receive continuing attention in the forests of the United States.

As we begin to contemplate potential global scale changes in our environment due perhaps to adverse and additional anthropogenic influences (250, 397), we must remember to account for normal biological/pathological phenomena (295). If cause/effect interactions are considered feasible (e.g., a change in suscept resistance or biotic pathogen aggressiveness), then research must be carried to full completion in order to irrefutably demonstrate such interactions. Inference, especially repeated inference, does not provide factual evidence of cause/effect relationships.

For reasons provided within this paper, forest biologists must take into account the total complex of organisms, their metabolic relationships, abiotic stress factors and their potential interactions when assessing altered tree or forest health. Pictorial atlases which describe common diseases and insects in European and eastern United States forests have been recently published (173, 437). These and all other available support materials should be used to increase accuracy of the diagnostic process. To do anything less, prior to concluding cause(s) of forest (tree) declines as being due to air pollutants in any of several potential forms would be biologically and therefore ethically impoverished.

All scientists must guard against the insidious trap that Gardner (142) referred to in the following: "Sincere belief mixed with crafty deception is conspicuous by its prevalence in the history of bogus science." We must be wary of "partial" or "directed" statements and crafty efforts to ignore additional yet important diagnostic information so as not to become part of the group identified by Maddox (287) who wrote:

"The environmentalists may be the most insidious of all plunderers of our planet. Using 'a technique of calculated overdramatization', they have deflected attention from the genuine ecological issues we face and blinded us to solutions that exist now."

Portions of this chapter have been published in S. K. Majumbar, E. W. Miller, and J. Cahir, eds., 1991, Air Pollution: Environmental Issues and Health Effects, Pennsylvania Academy of Science, Easton. Used by permission.

ALASKA YELLOW-CEDAR DECLINE: DISTRIBUTION, EPIDEMIOLOGY, AND ETIOLOGY

Paul E. Hennon[1], Charles G. Shaw III[2], and Everett M. Hansen[3]

[1] USDA Forest Service, Juneau, AK
[2] USDA Forest Service, Fort Collins, CO
[3] Oregon State University, Corvallis, OR

Decline and mortality of Alaska yellow-cedar, Chamaecyparis nootkatensis (D. Don) Spach, is one of the most spectacular and curious forest problems in Alaska. The problem now covers over 200,000 hectares of remote, unmanaged forest and is distributed throughout most of southeast Alaska. Symptoms of dying Alaska yellow-cedar trees are typical of other forest declines. They include dying fine roots, necrotic lesions on roots and boles, reduced radial growth, and foliage death distributed throughout the crown. The cause of decline has been variously attributed to bark beetles, root disease, and winter injury, but these suggestions were based on brief observations. Decline did not receive detailed investigation until the early 1980's when our studies began. Through studies of symptomology, associated biotic factors, and epidemiology (onset and development of decline), we have attempted to test the hypothesis that some biotic factor is the primary cause of cedar decline. The objective of this paper is to summarize the results from these various studies.

Cedar, the Victim

Alaska yellow-cedar, the principal victim of decline, is

a slow-growing conifer that ranges from Prince William Sound in Alaska, south through British Columbia, to near the Oregon-California border (169). It occurs from sea level to timberline in southeast Alaska (170) where it can grow in nearly pure stands but, more commonly, exists in scattered groups or as individual trees mixed with other conifers (388). Alaska yellow-cedar is among conifers with great longevity; trees over 1000 years old are common (131). Its narrow grain, extreme decay resistance, and bright-yellow, aromatic heartwood make Alaska yellow-cedar a useful and valuable timber species (133). On a per unit volume basis, Alaska yellow-cedar is the most valuable wood grown in Alaska.

Southeast Alaska, the Setting

Southeast Alaska has a cool, wet climate with an annual precipitation ranging from 150 to 500 cm (171). Winters have relatively moderate temperatures although brief cold periods occur. Summers are cool and wet, without prolonged dry periods; thus, fire is not an important factor in forest succession (170). Windthrow and landslides are common disturbances of forests in this region (170); mortality caused by diseases and insects also may be important disturbance factors. Patterns of plant succession on poorly drained sites are not understood. Poorly drained soils, which are highly organic and shallow or deep, generally occur on sites without steep slope and overlay unfractured bedrock or compacted glacial till (298).

This region is dominated by undisturbed, old-growth forests of western hemlock (Tsuga heterophylla (Raf.) (Sarg.) and Sitka spruce (Picea sitchensis (Bong.) (Carr.) on well-drained soils and by Alaska yellow-cedar, western hemlock, mountain hemlock (Tsuga mertensiana (Bong.) (Carr.), and shore pine (Pinus contorta var. contorta Doug.) on poorly-drained, deeply-organic soils (171). On extremely boggy sites, shore pine and prostrate Alaska yellow-cedar are the only conifers.

DISTRIBUTION OF DECLINE

The distribution of cedar decline portrayed in Figure 1 is based on detailed sketchings made on 1:250,000 scale maps during aerial surveys conducted annually during the 1980's.

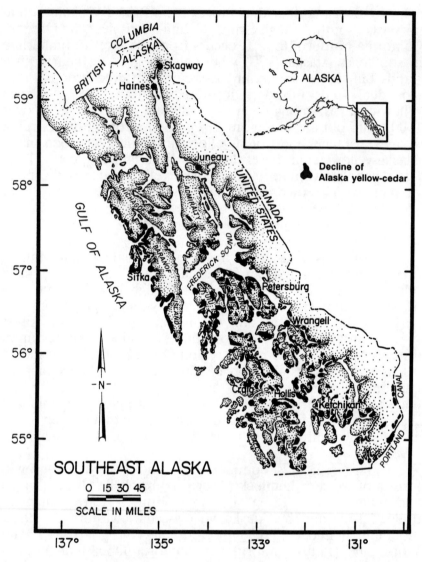

Figure 1. Distribution of severe decline and mortality (dark areas) of Alaska yellow-cedar in southeast Alaska. (The small, speckled dots do not represent decline; they delineate the land mass.)

Some 200,000 ha of decline have been mapped in southeast Alaska from the British Columbia border near Portland Canal to the northwest side of Chicagof Island. The area of severe decline occurs in a broad band from northwest to southeast. Decline is either absent or not severe in other portions of the natural range of Alaska yellow-cedar further south in British Columbia and around the Gulf of Alaska near the northwest limit of its range.

EPIDEMIOLOGY

Aerial Photographs and Ground Surveys

Analysis of aerial photographs and numerous ground surveys were conducted to determine in which forest communities decline occurs, which tree species are dying, and if mortality has been spreading over short or long distances. Such information should aid in evaluating potential causal factors.

Ground surveys were conducted in areas of severe forest decline on Baranof and Chicagof Islands in Southeast Alaska (182). Surveys were also conducted on Prince of Wales and Wrangell Islands near the northern limit of the distribution of western redcedar (Thuja plicata Donn ex D. Donn) to determine if this species suffers from decline.

Each dead Alaska yellow-cedar encountered on ground surveys was placed into one of six snag classes based on degrees of foliage, twig, or branch retention, and deterioration of its bole. Average time since death is estimated for five of these snag classes (184):

Class   I -- dead, foliage retained
........................dead 4 years
Class   II -- dead, twigs retained
..................dead 14 years
Class III -- dead, secondary branches retained
....................dead 26 years
Class IV -- dead, primary branches retained
.......................dead 51 years
Class   V -- dead, no branches retained
......................dead 81 years
Class VI -- dead, bole broken and deteriorated
...................death not dated

Understory vegetation was recorded during ground surveys and used in an ordination analysis (detrended correspondence analyses) to indicate relationships of mortality and plant communities. The ordination that resulted suggests that plant communities are strongly influenced by soil moisture drainage (182).

Decline Associated with Poorly Drained Sites

Our aerial observations and examinations made from recent color aerial photographs demonstrate that cedar decline is strongly associated with forests that occur on poorly drained sites. In many areas, mortality occurs on the edges of nearly all open bogs. Some bogs extend, contiguously or in chains, for many kilometers along fairly flat terrain at lower elevations where trees can be found dying adjacent to beaches. Mortality is consistently associated with the edges of these bogs. Decline is also severe in large areas without open bogs, but here the understory flora and tree stature suggest moderately poor drainage. Within the general distribution of severe mortality, forested areas without concentrated mortality generally lack bogs and are dominated by high-volume hemlock forests. Some boggy areas with Alaska yellow-cedar but not decline, however, occur at elevations above about 200 m. In addition, data from ground surveys indicate that the incidence of mortality is significantly more common ($P = 0.05$) in bog communities and is progressively less common in communities with better drainage (182).

Tree Species Dying

Alaska yellow-cedar is the principal species that is dead or dying in stands with a high proportion of mortality, accounting for 74% of the dead basal area (182). Dead hemlock accounts for 16.8% of the mortality in declining stands, with other species only having negligible dead basal area.

Since Alaska yellow-cedar predominates in most declining stands, the percent basal area dead for each species provides a better measure of which species have been affected. Sixty-five percent of Alaska yellow-cedar basal area is dead in declining stands, nearly twice the percentage of any other species. Alaska yellow-cedar is also dying in disproportionate levels in stands with western redcedar. On Prince of Wales Island, 34% of the basal area of Alaska yellow-cedar was dead compared to 9% dead for redcedar; on Wrangell Island, 54% of Alaska yellow-cedar was dead, but only 3% of the western redcedar was dead.

Mortality Spread

Maps delineating the extent of mortality at seven sites made from aerial photographs taken in 1927, 1948, 1965, and 1976 clearly show that peripheral boundaries of mortality have expanded at all sites (182). In 1927, the mortality apparent on each site covered a large portion of the area where trees are now dead and dying; thus, local spread since 1927 accounts for only a small proportion of the total area of decline. Subsequent mortality has rarely extended more than 100 m beyond the 1927 boundary.

Results from ground surveys also indicate local spread at many, but not all, sites that were surveyed. In such stands, dying and recently-killed cedars often surround areas containing the old snags. This mortality spread, which commonly occurs upslope but in any cardinal direction, has been along the gradient from bog to better drainage, as evidenced by understory plant analysis (ordination). Snags with no limbs (class V) predominate in bog and semibog plant communities and plots with more recently killed snags (classes IV, III, II, and I) support progressively better-drained plant communities. Snags in the longest-dead class VI (deteriorating boles) are uncommon, are not associated with severe mortality, and are not confined to bogs, as are class V snags. Thus, spread of mortality within any one site has occurred as a slow advance along an established ecological gradient which is often related to slope. The common upslope spread of mortality results from mortality originating in bogs and semibogs and spreading upslope along the gradient to communities with better drainage.

Ecological Effects of Decline

Different forest conditions have developed in the areas of earliest mortality where the long-dead (class V) snags are present. A new stand of vigorous-appearing trees has grown up beneath the bark-free, white snags on some sites. Alaska yellow-cedar, western hemlock, and to some extent mountain hemlock, are the dominant tree species in these areas and appear as a green zone from a distance or on color aerial photographs. Most or all of these trees are older than 100 years and were probably present as understory conifers during the initial mortality or, in the case of hemlock, are surviving overstory trees.

In other stands with long-dead snags, continued mortality of smaller Alaska yellow-cedar trees has apparently prevented development of this green zone. Reasons for the recurrence of mortality in some stands, but not in others, are unclear. On extremely boggy sites, release of live trees following mortality has not occurred, likely because factors suppressing the growth of live trees (e.g., anaerobic soils) are not improved by the death of Alaska yellow-cedars. Once the dominant overstory of Alaska yellow-cedar dies, soils may become wetter, as the reduced transpiration causes a degeneration of the site that may affect survival or growth of other conifer species. Interestingly, southeast Alaskan tree species suspected of being relatively intolerant to excessive moisture (388) suffer higher rates of death (35% of its basal area for Sitka spruce and 29% for hemlock) than have the tolerant shore pine (6%) on sites of mortality. Alaska yellow-cedar is the exception; the species is reportedly well adapted to wet sites, but it suffers the greatest mortality (65%) in declining stands.

Alaska yellow-cedar is not frequently regenerating from seed in declining or healthy forests. While collecting the data for understory plants in our surveys, we counted the number of Alaska yellow-cedar seedlings. Seedlings are never common; most occur in stands with the greatest live cedar basal area. Nearly all seedlings are in the germling stage, but do not become saplings. Alaska yellow-cedar is commonly regenerating by asexual means on wet sites where lower branches in contact with the ground root adventitiously. Once rooted, these lower branches sometimes detach and become separate plants and may slowly grow into trees. This asexual form of reproduction is common on sites with decline.

Snags lacking limbs (class V) were present on all sites with dead Alaska yellow-cedars and constituted at least 8% (range = 8-60%) of all snags on 23 sites with heavy mortality. More recently killed snags (i.e., classes I-IV) were also present along all transects, indicating that the mortality has continued at all locations since initiation. Our general reconnaissance revealed only one site that has only long-dead, class IV and V snags and lacks recent mortality. No sites, however, have recently killed trees in the absence of long-dead class V snags. Thus, we have no evidence of site-to-site spread of decline.

Dating the Onset

To determine how long cedar decline has been occurring in Alaska, we examined old aerial photographs, used two methods to date the death of standing dead trees, and inspected historical accounts of botanical expeditions (184).

The earliest available aerial photographs of southeast Alaska, taken in 1926 and 1927 by the U.S. Navy, represent one of the first efforts anywhere to photograph large areas of forest (391). These photographs are now of variable quality; but, on both vertical and oblique prints with good contrast, cedar mortality clearly appears as patches of white snags. Mortality of Alaska yellow-cedar was already widespread by 1927.

Estimates of the time since death for Alaska yellow-cedar trees in the snag classes were determined by counting annual rings of hemlock trees growing under large cedar snags and by counting annual rings in callus growth on partially killed stems of Alaska yellow-cedars (we call these rope trees) that were interspersed among cedar snags (184).

Of the 73 hemlocks examined growing beneath dead cedars, 58 (79%) released from their otherwise slow but relatively constant rate of annual growth (184). The times of these growth releases differed significantly among snag classes (I-IV) under which the hemlocks grew. Few hemlocks released under class V snags.

Rope trees have a dead top (snag class I to V) and one narrow strip of live tissue, consisting of callusing bark and sapwood, that connects roots to one live and bushy branch cluster. The cause of this condition is not known; however, we hypothesize that these trees were severely injured, but not completely killed, by whatever caused nearby cedars to die. Most rope trees have dead tops in snag class IV or class V; few rope trees have class I tops. Rope trees in classes II, III, and V differ significantly from one another in number of callus growth rings and, presumably, in time since death of the top and most of the bole (184). Rope trees were not injured during one sudden incident such as an extreme climatic event; the cambiums among the trees that we sampled did not die during the same year. Also, their slow decline in growth many years prior to bole death, initiated at different times, does not support a sudden event as the cause of tree injury.

We consider that class V snags (boles intact, but no primary limbs retained) are the original extensive mortality. These trees died an average of 81 years ago (184), as estimated by the rope tree method, and were present and common at all mortality sites examined. Because numerous class V snags died before the average of 81 years ago, some Alaska yellow-cedars probably began to die before the turn of the century, our estimate being around 1880. The early aerial photographs confirm the widespread occurrence of dead trees in 1927. The more deteriorated snags in class VI with broken-off and decayed boles were only infrequently encountered in surveys and were not associated with distinct mortality sites (182). These latter trees probably died prior to the onset of extensive mortality and may represent the nonepidemic or background level of mortality.

The appearance of abundant dead Alaska yellow-cedars around 1900 is also supported by historical observations. Sheldon (420) was the first observer to report extensive mortality and noted dead Alaska yellow-cedar near Pybus Bay on Admiralty Island in 1909, stating that, "vast areas are rolling swamp, with yellow-cedars, mostly dead". Numerous botanical expeditions to Sitka and other areas in southeast Alaska where cedar decline is now extensive were conducted prior to 1880; most report the occurrence of Alaska yellow-cedar, but none mentions dead or dying trees (184).

The excessive level of cedar mortality on sites of decline (65% of the cedar has died in the last 100 years), time required to establish mature cedar trees (nearly all trees are well over 100 years old), and inadequate replacement of cedar by regeneration suggest that the population of Alaska yellow-cedar in these forests is diminishing and that intensive mortality could not have been an ongoing phenomenon for centuries.

ETIOLOGY

Symptoms of Decline

Symptoms of decline were studied by annual examinations of more than 250 cedar trees over 9 years and by systematic observations made on 62 root-excavated cedar trees (185, 418). Crowns of declining Alaska yellow-cedars die as a unit, quickly or slowly, rather than as isolated dying branches. In cedars that die quickly, all foliage dies concurrently, leaving a relatively full, but red or brown crown. Slowly declining cedars have dead proximal (older) foliage that abscisses before the younger distal foliage dies which produces a crown with a thin appearance. Some such trees that we observed in an advanced stage of decline in 1981 were still living in 1990. Symptom development is progressive, even if prolonged, while recovery of declining trees is rare. The crown symptomatology suggests a below ground, root-related, problem.

Trees in early stages of crown decline have dead and missing fine roots, as well as root lesions that spread from small distal roots into and along larger proximal roots to the root collar. In the final stages of decline, vertical lesions spread from dead roots up the tree's bole. If a pathogen is primarily responsible for killing these trees, then it is likely to be located in one or more of these necrotic tissues. Vertical bole lesions do not, however, appear to be caused by fungal activity as no single fungus was consistently isolated from them. Implant inoculations with pieces of healthy bark caused as many lesions as those using necrotic bark taken from lesions. Lesions induced by wounding roots appeared similar to natural lesions and were colonized by a similar array of fungi.

Lesions on larger roots and boles of declining cedars are oriented primarily along the root axis or vertically on boles; they do not spread circumferentially. This non-girdling pattern of development suggests that these lesions do not directly result from activity by pathogens, but are secondary symptoms. Lesions apparently result from death of smaller roots, which occurs earlier in tree decline.

Biotic Factors Associated with Decline

Several thousand isolations for fungi were attempted from the symptomatic tissues described above (185). None of the 50 species of fungi (180) that we isolated or collected from Alaska yellow-cedar can be considered to be the primary incitant of the extensive mortality. Most of the fungi that were consistently collected from dying cedars were also found on healthy Alaska yellow-cedars. In inoculation trials with the 11 fungi most commonly isolated from symptomatic tissues, only Cylindrocarpon didymum (Hartig) Wollenweb. showed pathogenicity; it caused necrotic lesions, but failed to kill any cedar seedlings.

Mycelium radicis atrovirens Melin was isolated 235 times from fine roots and other tissues, and is most likely the dark-colored fungus observed in the cortical cells on fine roots of 79% of the Alaska yellow-cedars that we examined (185). Although M. radicis atrovirens has been repeatedly isolated from roots of various forest trees and is reportedly ubiquitous in boreal forest soils (140), the nature of its parasitism on conifers is unclear. The association with healthy cedars and its lack of pathogenicity in our studies suggest that M. radicis atrovirens is not a primary pathogen on cedar.

In our inoculation tests, Armillaria sp., which is common on dying and recently dead cedars (185, 418), neither initiated lesions nor did it kill roots on 16 mature cedar trees. In another study (419), isolates of Armillaria sp. from dying Alaska yellow-cedars failed to infect seedlings of cedar even though the inoculum survived and readily produced rhizomorphs. This Armillaria species appears to be a common, but distinctly secondary pathogen, on declining cedars. It colonizes roots that are already dead or dying, and occasionally hastens the death of declining trees by killing the cambium at the root collar, a behavior typical of its role in many other forest settings.

Ironically, members of both Phytophthora and Seiridium, genera known to cause serious diseases to Chamaecyparis elsewhere, were found associated with Alaska yellow-cedar in southeast Alaska. However, both were rare and probably cause little or no damage to cedar. Seiridium cardinale (W. Wagener) Sutton and I. Gibson was isolated only once, from a bear-caused scar. This species varies in its pathogenicity from an aggressive pathogen to a saprobe (481); the isolate from Alaska yellow-cedar lacked pathogenicity in inoculation studies.

A species of Phytophthora was isolated 4 times from baits around 69 cedar trees. It was never isolated directly from cedar tissues, even though selective medium was used in several hundred attempts (185). The fungus, which we recently identified as P. gonapodyides (Petersen) Buisman., is probably not the cause of Alaska yellow-cedar decline. The same species, referred to as P. drechsleri-like (163) because of similar morphology and electrophoretic protein patterns, was recovered from undisturbed watersheds that lacked significant tree mortality in Oregon (162).

In recent sampling for Pythium spp. at six general locations throughout southeast Alaska, including sites with extreme decline, five species were commonly recovered, but their recovery rate was not associated with mortality sites or soil samples from beneath dying trees (159).

The lack or dysfunction of beneficial mycorrhizae does not appear to contribute to Alaska yellow-cedar decline. Vesicular-arbuscular mycorrhizae were common in the cortical cells of live fine roots on both declining and healthy Alaska yellow-cedars (185).

Root-feeding nematodes (Pratylenchus sp., Aphelenchoides sp., Sphaeronema sp., and Crossonema sp.) were recovered from beneath cedar trees, but they were neither prevalent enough nor sufficiently restricted to sites of dying cedars to be primary inciters of Alaska yellow-cedar decline (183). Bark beetles (Phloeosinus spp.), which were once thought to be the primary cause of tree death, occur on a small proportion of dying trees in most areas and are restricted to attacking cedars in later stages of crown decline (133, 418).

Basal scars are common on many cedar trees in declining stands. Nearly one-half (49%) of the cedars sampled in the Peril Strait and Slocum Arm area have either fresh or old, callusing scars (181). Fresh scars consistently have teeth or bite marks and are almost certainly the result of feeding by Alaskan brown bears (Ursus arctos) (181). A smaller number of scars are caused by Alaska Native people stripping bark from cedar trees (181).

Regardless of cause, basal scars are not the primary cause of cedar decline. Scar incidence is not greater on dying cedar trees than on healthy trees, nor is it greater on cedars in declining stands than in healthy stands (181). Numerous Alaska yellow-cedar trees die with no basal scars (185). Basal scarring is most common in the well-drained forest type (181) where mortality is least common (182). Extensive mortality of

Alaska yellow-cedar occurs on Prince of Wales, Wrangell, and other islands where brown bears are absent and basal scarring was not found (181).

Possible Abiotic Causes of Decline

One hypothesis for an abiotic cause of tree death is that bogs, for climatic or other reasons, are advancing onto adjacent, semi-bog sites where so many trees are dying (257). The development from forest to bog requires the waterlogging of the forest floor, which may result from the proliferation of Sphagnum moss, the development of poor drainage, or both. This process, paludification, may lead to the death of forest trees as sufficient oxygen or nutrients become less available in the wet soil. Whether there is a general successional direction for forests in southeast Alaska, from forest to bog or bog to forest, is presently unresolved.

Our observations, however, suggest that if bogs are advancing on forests, then the rate of advancement is imperceptible. Bogs observable on the 1927 aerial photography have not noticeably expanded, yet cedar mortality has been substantial during this period. One might expect to see some evidence of rapidly expanding sphagnum mats or invasion of other bog plants into forests if bogs have enlarged during the last 100 years. In recent observations of understory flora growing beneath several hundred dying cedars at four sites in southeast Alaska, more than 75% of these trees had no Sphagnum spp. within a 1 m radius of tree boles (Hennon, unpublished data). Even if bogs are expanding, the rate of expansion is probably too slow to provide a simple explanation for the widespread decline of Alaska yellow-cedar. Also, the relatively high rate of mortality for Alaska yellow-cedar, perhaps one of the conifers best adapted to growing in bogs (aside from shore pine), compared to a lower incidence of mortality for other conifers, contradicts the simple hypothesis of expanding bogs killing cedars.

Soil toxicity, which presumably could affect Alaska yellow-cedar more than other conifers, may explain tree death. Perhaps toxic, organic compounds result from anaerobic decomposition in these wet, highly organic soils.

Another hypothetical explanation for cedars dying around bogs and on wet sites is the poor protection from atmospheric events that these sites offer. Cedar trees on such sites are open-grown and probably more vulnerable to extreme weather events (e.g., freeze, desiccation) compared to cedars grown within protective canopies. Conceivably, if some trees on bog edges die following such a hypothetical weather event, then trees in adjacent forest stands could lose protection and become vulnerable to damage. Such action might lead to a slow, local spreading of mortality from the open stands in bogs and semi-bogs.

Subtle variations in climate may be responsible for triggering some forest declines, such as those of maple (Acer), birch (Betula), and ash (Fraxinus). Since climate is not static, vegetation may be constantly adjusting to reflect its new environment. The long-lived Alaska yellow-cedar may have difficulty adapting to a changing environment, particularly if cedar is reproducing asexually by vegetative layering (179). Interestingly, a warming trend has occurred in most of Alaska since the late 1800's (158) which coincides with the onset of Alaska yellow-cedar decline. A slight increase in average winter temperatures would change some precipitation from snow to rain, reducing the snowpack at low elevations.

Perhaps the primary cause of cedar decline is freezing damage to root systems (fine root necrosis is apparently the initial symptom of dying cedars [185]) during periods when cold continental air moves over the region and roots are inadequately protected by snow. Trees growing in wet soils at low elevations, where decline has been severe, would be susceptible due to their shallow root system and poor insulation offered by saturated soils. Less decline on wet sites at higher elevations could be explained by the persistent winter snowpack at those elevations even in today's warmer climate.

SUMMARY

The cumulative results from these studies do not support the hypothesis that an organism is the primary cause of mortality. The specificity of mortality to Alaska yellow-cedar and evidence of local spread seem to suggest a pathogen-caused disease; however, no new sites of mortality appear to have developed since the nearly simultaneous onset of Alaska yellow-cedar decline some 100 years ago at numerous locations throughout southeast Alaska. Snags that represent the original extensive mortality were present at every sampling location and observed on all good-quality 1927 aerial photographs of sites where cedars are currently dying. It is difficult to imagine a pathogen capable of inciting and continuing to cause the level of destruction that occurs on remote and dispersed islands in isolated wilderness, but not capable of re-initiating the problem on other, similar bog and semi-bog sites. The pattern of local spread, along a preexisting physical gradient, also suggests that a pathogen is not involved.

Pathogenic fungi, nematodes, insects, and brown bears can, at best, only be concluded to have secondary or contributing roles in Alaska yellow-cedar decline. It now seems most likely that the primary cause of Alaska yellow-cedar decline is some abiotic factor(s).

Although the primary stress(es) in Alaska yellow-cedar decline is unknown, future research on etiology should focus on abiotic factors, such as properties of wet, organic soils (e.g., soil toxins from anaerobic decomposition) or atmospheric conditions that would affect cedars in bog and semi-bog communities where trees are poorly protected.

Alaska yellow-cedar decline appears to be a unique and outstanding example of a naturally induced forest decline. The extreme decay resistance of this species has allowed us the rare opportunity to reconstruct the onset and development of decline. The occurrence of extensive mortality before 1900 in countless remote, undisturbed sites without nearby sources of pollution argues against atmospheric pollution as the cause of decline. Additionally, no introduced exotic pathogen or insect (another potential form of anthropogenic activity) was found to be associated with decline. Whether or not a general warming trend, has affected the onset and development of decline, is an open question. Except for the probable abiotic cause, the specific, primary stresses in the etiology of Alaska yellow-cedar decline remain a mystery.

SUGAR MAPLE DECLINES - CAUSES, EFFECTS, AND RECOMMENDATIONS

Douglas C. Allen[1], Eric Bauce[2] and Charles J. Barnett[3]

[1]State University of New York College of Environmental Science and Forestry, Syracuse, NY 13210-2778
[2]Laval University, Sainte-Foy, Quebec, Canada G1V 4C7
[3]USDA Forest Service, Radnor, PA 19087-4585

Much was written about sugar maple (Acer saccharum Marsh.) health in both the scientific and popular press during the 1980's. These articles were prompted by apparent widespread signs of deterioration of maple across the range of northern hardwoods. Among the published materials, comprehensive literature reviews about major northern hardwood insect pests and diseases (3, 212, 215) and sugar maple crown dieback/tree decline (40, 308, 319) focused attention on a variety of potentially harmful organisms, detrimental forest management and utilization practices, adverse weather events, and suspect forms of atmospheric pollution. All have been implicated as potential contributors to the deterioration in sugar maple that landowners apparently have observed at many diverse locations.

This widespread concern for sugar maple during the past decade focused much attention on the health of an important tree species. Yet the speculation and frequent sensationalism inherent in many recent portrayals of maple health have also handicapped forest health specialists and forest managers. In particular, many landowners question the credibility of these people who take exception to simplistic explanations of maple decline or crown dieback as commonly proclaimed in the media. This has hindered opportunities to look for possible complexes of causal agents, to identify the widest possible array of contributing factors, and to weigh them objectively by the normal process of scientific scrutiny.

The assessment presented here is based on four historical examples and one case study of sugar maple declines. It also reflects discussions with managers of northern hardwood stands and sugarbushes. Both sources are used as a basis for forest management recommendations. We also have identified some biotic and abiotic agents, often tempered by direct human intervention, that appear to play important roles in maple declines. In many cases, our assessment draws upon circumstantial evidence because clear causes cannot be demonstrated. Yet we feel that several suspected events have been associated frequently and consistently enough with crown dieback and maple decline to provide a basis for several useful forest management recommendations. We present them for consideration, and encourage their use as interim measures pending more definitive information that may eventually emerge from long-term research into maple crown dieback and tree decline.

Sugar maple is one of the most important broadleaved trees in the northeastern United States and southeastern Canada. Its continued ecological dominance in most of these communities is likely insured due to its high seed production, high shade tolerance, and long life. The income attached to this northern hardwood is especially important to the economic health of many rural areas. The value accrues either individually or collectively from stumpage sales, primary processing, secondary manufacturing, the maple syrup industry, and tourism. Employment opportunities and cash flow are limited in most of these rural areas, and the diverse array of maple products and tourist benefits help to ease both problems. We believe that these ecological and economic contributions justify stronger research support to address the health and management of northern hardwood stands. They certainly attest to the importance of sugar maple throughout this region.

WHAT IS DECLINE?

Declines have been defined in different ways. These divergent interpretations often confuse both laymen and scientists, which obstructs understanding the disease. Our view of a tree decline is based on the "decline spiral" concept proposed by Manion (291, 294), and the dieback/decline philosophy contributed largely by Houston (207, 209). These two schools of thought have much in common, and basically differ only in the way they categorize different stress factors. Manion's "spiral" is based, in part, on the three-tiered hierarchy of factors associated with declines that was proposed originally by Sinclair (429) (predisposing, inciting, contributing). Houston, on the other hand, placed stress in one of two categories (preliminary stress, secondary events). Two important aspects of a tree decline are: i) it is a disease resulting from a complex of stresses acting in succession or contemporaneously, and ii) it is characterized by a sequence of events, often including the ultimate demise of the tree, which operate over a period of several years.

Crown dieback, though frequently symptomatic of a decline, also can signal a temporary response to isolated, short-term stresses. One-time, solitary events such as drought, defoliation by insects, certain diseases, physical damage to the tree bole, or a severe late spring frost can precipitate dieback. Yet in these instances, the tree crown usually recovers in two or three growing seasons, as long as the disturbance ameliorates and does not trigger a series of subsequent stresses. Following the short-term scenario, dieback is a temporary response that reduces demand for energy to produce foliage and branches (207). Whether or not the crown recovers, or the initial stress initiates a sequence of events that triggers the disease called decline, depends on stand disturbance history (i.e., the condition of the tree at the time of stress) and subsequent biotic and/or abiotic stresses. These may include forest management activities.

CAUSES OF SUGAR MAPLE DECLINE - HISTORICAL EXAMPLES

During the twentieth century, several biotic and abiotic stresses have been associated with sugar maple "declines" in forest stands and sugarbushes. Events summarized by McIlveen et al. (308) and Millers et al. (319) indicate that the agents responsible for many episodes of "decline" are unknown. Since the presence of dieback does not, a priori, signify a decline, it is unclear whether many examples presented in these two reviews represent legitimate declines. In fact, few historical records and/or little empirical evidence are available to verify the occurrence of or explain the etiology of a maple decline. By contrast, we have identified four declines that have been documented through repeated field observations or as a result of detailed investigation. These illustrate two important characteristics of most sugar maple declines: i) the variety and nature of stresses normally involved; and ii) the significant role of human activities. Additionally, they illustrate the important role played by on-site events.

The study of maple blight in Florence County, Wisconsin during the late 1950's and early 1960's (6) is the most thoroughly documented investigation of a maple decline. In this instance, combined effects of defoliation by a complex of insects, infection by Armillaria root rot, frost damage, low stand density, a high proportion of stocking in sugar maple, and reduced soil moisture were implicated in the genesis or development of the disease (5). Two waves of selective timber harvesting early in the century created a stand dominated by sugar maple, and probably initiated the sequence of events that eventually led to maple decline (151).

Observations in Vermont's northern hardwood forests from 1977-1985 associated deteriorating sugar maple with the frequency, level, and timing of defoliation by forest tent caterpillar (Malacosoma disstria Hübner); frost damage to foliage; and occurrence of stand thinning immediately prior to or during the insect outbreak (246). Additionally, losses appeared highest in stands that were predominantly (90%+) sugar maple and poorly managed (460).

Sugar maple dieback that progressed into an apparent decline first appeared in northwestern Pennsylvania in the early 1980's. The affected areas (totalling approx. 1,214 ha) were restricted to ridge tops (above 640 m) and the symptoms coincided closely with episodes of insect defoliation during the previous 20 years (466). The sites were considered marginal for sugar maple. They also occurred on relatively level terrain that made them attractive for repeated logging (427).

Recent study of a maple decline by Bauce and Allen (47) implicated excessive stand density as a predisposing agent, drought, and low winter temperatures accompanied by lack of snow cover as inciting factors, and sugar maple borer (Glycobius speciosus Say) and Armillaria root disease as contributing agents. The stand studied was principally sugar maple (80% of basal area), and it had not been managed.

A CASE STUDY OF SUGAR MAPLE DECLINE: EFFECTS ON SAP PRODUCTION AND TREE GROWTH

Long-term information about ecological conditions in a stand is essential to document influences of various disturbance events on plant growth and community development. The most useful information is provided by sustained ecological monitoring. While the difficulties of maintaining such programs are many (e.g., 130, 450), effects of events like outbreaks of insects and diseases can be misleading if viewed solely in the short term (e.g., 209, 506). Time lags between causes and effects of ecological disturbances are to be expected. Rarely can the net effect be determined when events are viewed solely in a short time frame. When evaluations of stress events ignore the temporal influence, serious misjudgment and, consequently, mismanagement are likely (289).

Documentation of pre-outbreak stand conditions, knowledge of forest management history, and observations on the extent of previous natural disturbances are also required in order to interpret the real potential ecological and economic impacts of such outbreaks.

The stimulus to initiate long-term monitoring of sugar maple growth and sap quality occurred following a major outbreak of saddled prominent, Heterocampa guttivitta (Walker) (1967-1973). At its peak in 1969, the damage encompassed more than 445,170 ha in New York and New England (155). During and immediately following extensive defoliation, the effects on sugar maple mortality, growth, and sap production were documented in several regions. However, these data, coupled with observations made during a subsequent outbreak in 1981 (3), showed effects of the defoliation in only a narrow time frame, and included only limited information about stand management or disturbance history. To broaden the scope, monitoring was extended at a northern New York site to include sugar maple health (crown condition), diameter growth, and sap quality, and to characterize changes in tree and stand conditions over time. It was believed that access to such background information (stand and individual tree histories) would permit a more accurate assessment of impact when any future disturbance occurred in the stand.

The project reported here began in 1975 at the Dubuar

Forest of the State University of New York, College of Environmental Science and Forestry (Lat. N44°09', Long. W74°54'). The selected sugarbush appeared healthy; trees had full crowns with leaves of normal size and color. Crown dieback first appeared in 1978 and by 1981 14% of a sample of dominant and codominant sugar maple trees (n = 347) showed evidence of crown dieback. By 1986, 23% of these trees had dieback or were in decline, and 22% were dead. Based upon these data, the problem was classified as a decline.

Study Description

Sugar maple comprised 80% of the basal area (25.3 m^2/ha) in the 76 year-old (in 1975), even-aged northern hardwood stand under study. Red maple (A. rubrum L.), black cherry (Prunus serotina Ehrh.), and American beech (Fagus grandifolia Ehrh.) were the principle dominant and codominant associates. The cover type corresponds most closely to SAF type 25 (121).

In 1977, sample trees were selected at random from a population of dominant and codominant sugar maples that comprised this operating sugarbush. They were grouped into three plots to represent west, central, and east portions of the sugarbush. The western and central plots are approximately 100 m apart, and the eastern plot is 0.8 km from the central plot. Initial average diameter at breast height (d.b.h.-1.4m) was 13.3±1.7 cm (n = 13), 13.7±1.5 cm (n = 22), and 13.7±1.3 cm (n = 20) for the east, central, and west plots, respectively. During the study, three of the original 64 sample trees died, and dendrometer bands were damaged or misread on another six.

Dendrometer bands were installed on each tree (278) at 1.4 m (breast ht.) and 5.3 m (top of first log and usually base of the crown). Initial reading (zero point) of each band occurred in March, 1978. Bands were read each March prior to bud swell, and annual growth was obtained by converting diameter readings to cross-sectional area of wood produced from the end of one growing season to the beginning of the next. All trees were tapped annually beginning in 1978 using standard practices (512). Five to ten measurements of sap sugar content were taken annually from each tree using a hand-held refractometer (438). Every year, during late July to early August, the condition of each tree crown was evaluated for defoliation and dieback using the dieback classes listed in Table 1. Seed production was also determined as follows: no or very little seed, moderate seed production (seeds obvious, but not abundant enough to alter drastically crown appearance); and heavy seed production (branches in upper crown reddish to light brown). No significant insect defoliation occurred during the period of observation. However, the stand has experienced two defoliator outbreaks; forest tent caterpillar in 1953 and 1954 and saddled prominent in 1967 and 1968 (47).

The west plot was thinned in 1976 to allow for crown expansion of residual sugar maples. At this time, average basal area was reduced from 25.3 m^2/ha to 16.1 m^2/ha.

Table 1. Condition classes used to classify sugar maple crowns.

CLASS	DESCRIPTION
I	Tree apparently healthy.
II	Status questionable: crown appears abnormal (foliage clumped, dwarfed and/or off-color), but no evidence of dieback.
III	Incipient dieback: a few dead terminals present in uppermost 10% of crown, and foliage in uppermost margin of crown abnormal.
IV	Light dieback: 10-25% of uppermost crown with evidence of dieback and abnormal foliage.
V	Moderate dieback: >25% to <50% of upper crown with dieback, abnormal foliage, and epicormic branches.
VI	Severe dieback: $\geq$50% of upper crown involved with the above symptoms.
VII	Tree dead.

Originally, all sap was collected with buckets, but in 1985 plastic tubing was installed in the east plot. Heavy equipment is not used off the road that goes through the sugarbush.

Dieback and Mortality

When the 64 sample trees were selected and banded in 1977, all had apparently healthy crowns. With one exception, they showed no evidence of dieback prior to 1981, three years after it was first noticed in other parts of the stand. The one tree that had crown dieback in 1978, died in 1980. As of 1990, only two other sample trees have died. Symptoms in one case were first noted in 1981, and the tree died by 1984. The entire crown of the second tree was sparse (small leaves) in 1985, and the tree died in 1987. The uniform deterioration of the crown, the rapid death, and the characteristic stain in the buttress roots indicated that the latter probably succumbed to sapstreak disease (252).

Of the remaining 61 trees, 13 have shown progressive crown dieback since 1981 or 1982. Additionally, dieback was first recorded for single trees in 1984, 1985, and 1988. Therefore, 16 of the remaining sample trees currently have evidence of significant dieback (i.e., rated Class IV or higher, see Table 1). Five trees at some point in time had evidence of significant dieback, but recovered. That is, crown ratings showed no or progressively less evidence of dieback in succeeding years. Nine of the 16 trees with crown dieback have damaged lower boles (sugar maple borer - 6, physical damage - 1, butt rot - 1, seam - 1). Five (11%) of the 45 sample trees whose crowns remain ostensibly healthy also have damaged boles.

Sap Quality

In 7 of 11 years, average sugar content of declining trees (class IV and higher) was higher than for trees with no evidence of dieback (Fig. 1). However, this difference was significant ($p = .05$) only in 1984 ($2.5 \pm 0.09\%$ vs $2.3 \pm 0.07\%$). Another indication that trees under stress tended to have elevated sugar concentrations is indicated by the pattern of sugar content for the trees that died during the period of study. For the three years preceding death, the annual average sugar content of these individuals was 0.7% (range 0.1 - 1.8%)

higher than the annual average for their respective plots. Sample size varied as the study progressed, because for some trees death or crown dieback stopped or prevented sap flow.

Average sugar content of maple sap has changed little since 1981 (Fig. 1). Observations made when annual sap quality measurements were taken, however, indicated that sap flow (volume) from declining trees was reduced markedly. This variable was not measured, but it became more difficult each year to obtain sap quality measurements from declining trees, because of intermittent and slow sap flow. This seems reasonable, because sap volume is related to tree vigor and crown size (321, 322).

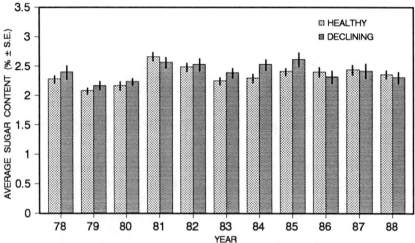

Fig. 1. Annual average sugar content of sap from apparently healthy (n=38) and declining (n=15, except n=14 in 1980, 1984 and 1987) sugar maples. Wanakena, NY.

Tree Growth

Average cross-sectional area (CSA) growth at breast height and top of the first log varied from 1972 through 1989, but generally decreased in all three plots during this period (Fig. 2). CSA growth at the top of the first log was significantly ($p = .05$) higher in the west plot (22.6 ± 1.4 cm^2) compared to sample trees in either the east (18.3 ± 2.4 cm^2) or central (17.3 ± 1.8 cm^2) plots. Similarly, average breast height growth (Fig. 2) for trees in the west plot (36.0 ± 2.4 cm^2) was significantly higher than in the east (25.8 ± 3.5 cm^2) or central (26.2 ± 2.4 cm^2) plots. During the last year of measurement (1989), growth at breast height was significantly lower in the central plot (10.4 ± 2.1 cm^2) compared to growth in the east (20.1 ± 3.1 cm^2) or west 19.4 ± 2.3 cm^2) plots.

Average annual growth at breast height for trees that had no evidence of crown dieback (i.e., were apparently healthy), was greater than growth of declining maples from 1979 through 1989 (Fig. 3). This growth difference was significant (p = .05) during three of these ten years (1985 through 1987). Periodic annual increment at breast height for the entire 12-year period was significantly greater for apparently healthy maples (21.7 ± 0.5 cm^2) compared to declining maples (17.2 ± 0.9 cm^2). Similarly, periodic annual increment at the top of the first log of healthy maples was significantly greater (13.7 ± 0.3 cm^2) than that of declining maples (10.7 ± 0.5 cm^2).

DISCUSSION

A recent study of decline in northern New York (47) used stem analyses to examine sugar maples in four classes of crown condition, ranging from apparently healthy (no dieback) to 50% or more crown dieback. Their review of stand management, weather events, and disturbance histories, indicated that excessive stand density and adverse climate (principally winter thaws and summer drought), played key roles in determining the present stand condition. Changes in growth since 1959 were significantly correlated (r = .95) with climatic events. Armillaria root rot and sugar maple borer were important secondary agents that apparently prohibited recovery of many affected trees.

The progressive deterioration of sugar maple crowns since 1981 as reported here for another Wanakena site also indicates a decline disease. In this case, only three of the 64 sample trees died during the period, but growth of all individuals has slowed since 1982 (Fig. 2).

The relatively good growth that occurred initially in the west plot probably was a response to thinning in 1976. Even though growth for all trees generally decreased during the study, only 16 of 61 trees currently exhibit crown dieback. Many of the latter have had evidence of dieback for the last nine years (1981-1989), yet remain alive.

The general response of both healthy and declining sugar maples to annual growing conditions was similar (Fig. 3).

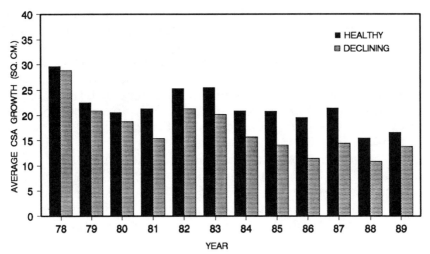

Fig. 2. Average annual cross-sectional area (CSA) growth at breast height for apparently healthy (n=38) and declining (n=15) sugar maples, 1978-1989. Wanakena, NY.

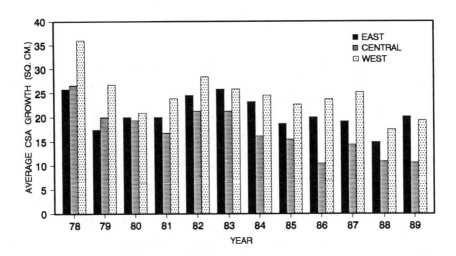

Fig. 3. Average annual cross-sectional area (CSA) growth at breast height for sugar maples in the east (n=11-13), central (n=23-24) and west (n=20-22).

After the disease appeared in 1980-1981, the pattern of growth for declining trees continued to parallel that of trees with healthy crowns but at lower increments.

Except for changes at d.b.h. in 1983 and 1989, annual sugar maple growth in the thinned plot (west) always exceeded that of maples in the other two plots after 1981 (Fig. 2), but the pattern of growth was similar for all plots. The parallel growth responses suggest that all plots were affected similarly by growing conditions during the past eight years. Following 1981, however, growth of trees in the thinned plot increased dramatically. This increased and sustained dominance may reflect the benefits of thinning. The stress of inter-tree competition is more intense in the other two plots, because they were not treated.

RECOMMENDATIONS

Our observations of several declining maple stands in New York during the past ten years and literature that provides experimental evidence or additional observations about maple declines, suggest that a forest manager who is aware of potentially threatening events can take the following steps to reduce susceptibility to decline. Among these, various kinds of natural and people-induced injuries will likely prove the most important to control. To this end, we suggest the following:

i) Whenever practical, discourage the development of stands that are made up of single-species and are even-aged. From the standpoint of both insects (e.g., sugar maple borer, maple leafcutter, forest tent caterpillar, maple webworm) or diseases (e.g., sapstreak, Eutypella canker), low species and structural (age class) diversity make a stand more susceptible to outbreaks of certain pests, and more vulnerable to damage.

ii) Pest detection and survey programs should be part of the forest management plan. By monitoring key pests, a forest manager can keep track of changes in populations or damage, thereby recognizing incipient conditions. Early recognition of changes in population or damage trends is necessary for planning and early implementation of prevention and control measures.

iii) If a recently stressed stand (e.g., drought, silvicultural treatment) is threatened by insect defoliation, a direct pest control measure should be considered to save foliage and minimize the impact of a potential inciting factor. Saving foliage also may determine whether or not a sugarbush is operable the following spring.

iv) Avoid thinning or other utilization (e.g., tapping) of stands that have been stressed recently by drought or heavy (50%+) insect defoliation. Waiting until the trees have recovered will reduce chances for aggravating the situation by imposing an additional stress.

v) When thinning is feasible, retain a basal area of approximately 16 m^2/ha (70 ft^2/acre) and minimize physical damage to the residual trees and soil. Observations suggest that thinning below this level predisposes sugar maple to exposure that may precipitate a decline. This seems especially true on sites with shallow soils, and in stands that have been highgraded repeatedly. Our observations on the potential negative effects of over-thinning parallel the views of others (e.g., 154, 251), who noted an association between dieback and over exposure of trees among heavily cut residual stands. This may be especially troublesome when trees are growing on sites with shallow soils or among stands recently stressed by defoliation or drought. Such stands often show stress within two years of the thinning.

AWARENESS AND PRUDENCE - BASIC INGREDIENTS FOR HEALTHY SUGAR MAPLE

Maple declines can frustrate landowners and foresters alike. Many key predisposing, inciting, or contributing factors are unpredictable or unmanageable. Stresses like drought, winter thaws, and late spring frosts are difficult to anticipate and impossible to prevent. Insect defoliation is predictable only when a monitoring system is in place to follow population change over time.

Silviculture is important in maintaining stand health and resiliency to many disturbances. This includes use of an appropriate reproduction method to assure a species mixture tolerant of local site and potential pest conditions, and regular and repeated tending of an age class as it matures. Yet lack of markets for small diameter and otherwise low quality logs and bolts prohibits implementation of desirable silvicultural programs in many northern hardwood stands. Indeed, market limitations encourage highgrading in the form of diameter-limit cuttings (344) that remove large diameter trees of superior genotypes, which may erode the overall genetic quality of a stand (306). Further, the temptation to highgrade northern hardwood stands will not likely subside in the near future. Sawlogs, particularly from shade intolerant species, will continue to bring higher stumpage prices, and short-term financial considerations will encourage their removal by many landowners. This will lead to stands dominated by shade-tolerant species, especially sugar maple. Widespread improvement in silvicultural practices is needed, particularly among the vast area of even-aged stands that comprise much of the eastern deciduous forest. Many of these 60 to 80 year-old, second-growth stands have not been thinned, are overstocked, and have a higher percentage of defective and high risk trees throughout the diameter distribution (141, 344). It is improbable, however, that these stands will receive treatment unless markets open for large quantities of low quality and small diameter wood. So, what does this portend for sugar maple health? Generally, we anticipate conditions will remain ripe for continued episodes of dieback and declines. The stress of intense inter-tree competition and pervasive low vigor that characterize many trees will predispose maple to a variety of debilitating events. The end result will be continued slow growth of all trees on many sites, and eventual mortality of some trees.

ACKNOWLEDGEMENTS

The comments of Paul D. Manion and Ralph P. Nyland, State University of New York, College of Environmental Science and Forestry, Syracuse contributed greatly to this manuscript. John W. Quimby, Bureau of Forestry, Div. of Forest Pest Management, Middletown, PA provided information regarding a maple decline in Pennsylvania.

A QUANTITATIVE TREE CROWN RATING SYSTEM FOR DECIDUOUS FOREST HEALTH SURVEYS: SOME RESULTS FOR ONTARIO

D. McLaughlin, W. Gizyn, W. McIlveen and
C. Kinch

Ontario Ministry of the Environment, 880 Bay Street, Toronto, Ontario Canada M5S 1Z8

Forest decline has been widely reported in popular and scientific literature in the past century (98). However, the frequency of these reports has increased dramatically in the last 10 years. These recent reports often cite atmospheric pollutants as contributing or even inciting agents in these decline phenomena.

West Germany reported decline of silver fir and Norway spruce in the early 1970s; and in the 1980s forest decline (of most coniferous species) was reported in Britain, Norway, Switzerland, Austria, France, Hungary, Czechoslovakia and East Germany (49, 65, 68, 76, 227, 469). In the U.S., coniferous forests are affected in the northeastern states and in parts of California (230, 415). In Canada, sugar maple decline has been historically reported across most of its range (308). Currently it is widespread in southern Quebec (57).

There are presently more than 200 theories on the causes of forest decline, which eludes to the complexity and contentiousness of the phenomenon (197). Regardless of the causes, all forest decline episodes have a common factor; that is, the trees develop characteristic visual symptoms of deterioration.

Individual trees have to be rated in some manner to arrive at a measure of tree condition, which can then be extrapolated to the stand level. Surveys of a number of stands are used as a measure of forest health at the local or regional scale. The most common approach, currently used by Sweden, West Germany and Great Britain, is a broad band assessment of crown density or defoliation; eg., $\leq$ 10%, 11%-25%, 26%-50%, etc. Another frequently used approach is a simplified numerical gradient; eg., 1 to 5 or 1 to 10, usually with the lowest number equivalent to a tree in the best condition.

These systems, although in wide use, have three significant shortcomings. First, by concentrating only on crown density (which by the lack of foliage implies twig/branch dieback) they are recognizing the last stages of the decline phenomenon, i.e., the dying tree. Therefore, their use as a warning of imminent decline is limited. Second, they tend to be qualitative, which substantially limits the application of descriptive statistics. The relatively low resolution of the assessed state of decline, using these methods, is the third shortcoming. A shift of one decline category can represent a significant change in the inferred tree condition. This is particularly critical in light of most quality assurance procedures which "allow" an assessment variation of plus or minus one category (i.e., on the 1 to 5 scale, a tree assessed as a 3 could possibly be a 2 or a 4). Cumulatively, these problems make subtle changes in tree condition difficult to measure confidently, thereby possibly obscuring trends.

The greatest shortcoming to the scientific community is the almost total lack of standardization of forest decline rating systems. This makes it impossible to directly compare forest health survey results between countries, and in many cases, between regions within the same country. Exceptions to this are recent efforts under way in Europe involving 18 countries (17), and the North American Sugar Maple Decline Program (215) among eastern Canadian provinces and northeastern states of the U.S.

This paper will describe a new, quantitative method of deciduous tree evaluation. The rationale behind the parameters selected for this method are discussed, as are the results of field trials to test the new technique. In addition, the results of three years of field data obtained from Forest Health Surveys in Ontario are discussed to illustrate a practical application of the assessment methodology.

A NEW, QUANTITATIVE FOREST HEALTH SURVEY TECHNIQUE

In Ontario, the symptoms most often associated with sugar maple decline are (312):

- delayed spring bud flush
- early fall leaf colour
- premature leaf abscission
- variation in foliar chlorosis
- undersized leaves
- progressive dieback of the fine twig structure
- progressive branch dieback
- epicormic sprouting
- reductions in annual radial xylem increment
- increased root mortality

These symptoms are indicative of stresses that cumulatively may result in tree death. Mortality may be quite sudden (two years) or it can take up to five or more growing seasons. Recovery of symptomatic trees has been documented. Of these symptoms, the most consistent and useful for purposes of forest health surveys, are estimates of leaf size, leaf colour and twig/branch dieback.

When the decision was made in 1984 to initiate deciduous forest decline studies in Ontario it was considered imperative that a rating system be developed that was quantitative, descriptive, reproducible, and had a high degree of resolution with a narrow confidence interval so that subtle differences could be detected. A high resolution, quantitative rating system was necessary so that regional differences in Ontario's hardwood forest could be recognized, and accurate detection of forest health trends facilitated. The method developed to rate the condition of Ontario deciduous forest trees was called the Decline Index (DI).

Field templates were prepared depicting tree crown silhouettes in 10% defoliated categories (Fig. 1). Sugar maple foliar samples were collected from numerous locations across the province. These samples were selected to represent a characteristic foliar colour gradient. The foliar samples were matched with colour paint chips representing six variations of three ranges of foliar colour: i) healthy/normal green, ii) slightly chlorotic, and iii) strongly chlorotic. The paint chips

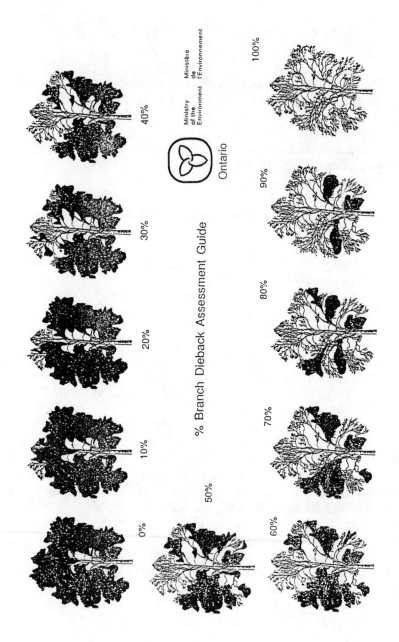

Fig. 1. Tree crown silhouettes used with the decline index field templates.

were mounted on the reverse side of the sheet of tree crown silhouettes and the template was laminated for durability in the field.

Of all the visible symptoms commonly associated with sugar maple decline in Ontario, twig/branch dieback and foliar colour and size were the three most frequently cited on symptomatic trees in a survey of more than 650 members of the Ontario Maple Syrup Producers Association (311). Therefore, these three crown parameters were chosen as the basis of the crown rating system. With the field templates as reference guides, field staff were trained to evaluate, on an individual tree basis, twig/branch dieback, foliar colour and the extent of dwarf foliage to the nearest 10% of the whole tree crown. The individual assessment parameters were recorded on a field tally sheet and the data were later transcribed to a spreadsheet file where the DI is automatically calculated. The three descriptive crown parameters (twig/branch dieback, foliar size and colour) were combined in a weighted formula which yields a numeric DI value ranging from 0 (healthy tree, no symptoms) to 100 (dead tree).

The DI formula is:

$$DI = DB + (A*UL) + (A*ST) + (A*SL/2)$$

where DI = decline index
DB = percent dead branches
UL = percent undersized leaves
ST = percent strong chlorosis
SL = percent slight chlorosis
A = a weighting factor (100-DB)/400

The weighting factor relates the foliar characteristics proportional to the live crown. Therefore, a tree which has an abundance of small and/or chlorotic leaves but relatively little twig/branch dieback will have an elevated DI. The formula is additive, with four separate components, therefore division of the final product by 400 was necessary to arrive at a value out of 100. Slight foliar chlorosis, most commonly described by maple syrup producers and woodlot owners as "distinctly off-green", was one of the most consistently reported symptoms in the maple syrup producers survey, yet it is likely the most difficult to quantify. Because it was such a consistent symptom its inclusion in the DI formula was important. As it would likely be the most difficult to assess confidently, it was decided (arbitrarily) to half the weighting value.

Foliar symptoms in the absence of significant twig/branch dieback are indicative of a short term stress and are often reversible. Persistent foliar symptoms usually indicate a chronic stress and may be followed in subsequent years by twig/branch dieback. Therefore, the DI should be able to identify subtle changes in tree condition associated with short term stress and should be useful as an early warning mechanism to identify pending tree decline. As with other rating systems, individual tree scores can be aggregated to infer the decline status of the stand or area.

HOW THE DI WORKS

To illustrate how the DI responds to a range of crown characteristics, data for nine hypothetical trees are summarized in Table 1. The example trees in this table have also been given subjective decline categories ranging from "healthy" to "severe". In these examples, the third and fourth trees have similar DIs (11 and 13, respectively) even though the third tree has no branch dieback. By comparison, the sixth and eighth trees have the same amount of branch dieback (20%) yet have very different DIs (25 and 50, respectively). In both cases, the higher percentages of foliar characteristics in the absence of

Table 1. Examples of hypothetical tree conditions to illustrate how the four parameters that make up the decline index respond to various crown characteristics.

Decline Category	% Branch Dieback	% Slight Chlorosis	% Strong Chlorosis	% Small Leaves	Decline Index
(1) Healthy	0	0	0	0	0
(2) Healthy	0	10	0	10	4
(3) Trace	0	30	10	20	11
(4) Trace	10	10	0	10	13
(5) Low	10	30	0	20	17
(6) Moderate	20	10	10	10	25
(7) Severe	30	10	20	30	40
(8) Severe	20	20	60	80	50
(9) Severe	60	20	40	90	74

branch dieback increases the resultant DI and illustrates the usefulness of the weighting factor. The DI method should be able to identify subtle annual changes in tree condition which would go unquantified by assessment methods that use only branch characteristics or simply the presence or absence of foliage.

Although it appeared that the DI method had the desired attributes for the planned deciduous forest health survey in Ontario, it was necessary to conduct field trials to determine if the rating method was reproducible and to establish confidence intervals. Ten field staff were trained in assessing tree decline symptoms. Ten trees were selected that represented a gradient of decline characteristics. To reduce assessment bias, the test trees were scattered throughout the test woodlot. Each evaluator assessed each tree five times. The tree numbers and order of evaluation were changed between assessment runs. A second set of five assessment runs was conducted on the ten test trees using paired evaluators.

A paired assessment means that two people discuss the condition of the same tree and arrive at a consensus regarding the score for each of the four decline-rating categories, and then fill out one evaluation sheet for that tree. As in the first set of assessments, the tree numbers and order of evaluation were changed between runs.

The results of the DI test trial are summarized in Table 2. Using trained evaluators with the assistance of standardized field templates, the DI method is clearly reproducible and it has an acceptably narrow confidence interval. This trial also revealed that paired assessment was usually better than a single person evaluation. The 99% confidence interval for the single-evaluator assessment ranged from 2.2 to 5.2, whereas the range was 1.4 to 4.5 for the paired assessment. Similarly, the coefficient of variation was lower for the paired assessment for all but the healthiest tree with the lowest DI. The coefficient of variation was inversely related to the DI. This is because the proportional difference in the DI is much greater for healthier trees relative to trees in a more advanced stage of decline. For example, a healthy tree with a DI of 4 may have a range of 2 to 6, which proportionately is much larger than a declining tree with a DI of 40 and a range of 35 to 45.

The DI system was designed to be used in the deciduous forests of south and south central Ontario, in which sugar

Table 2. Results of decline index reproducibility trials using single (one person) and paired (two people assessing each tree) evaluators.

Tree Number	Single Assessment*			Paired Assessment**		
	Mean DI	Coeff. of Var.	99% Con. Int.	Mean DI	Coeff. of Var.	99% Con. Int.
1	7	92%	2.2	2	165%	1.4
2	8	84%	2.2	8	49%	1.8
3	14	70%	3.2	13	35%	2.3
4	18	59%	3.6	17	54%	4.5
5	18	61%	3.8	21	38%	4.0
6	23	67%	5.2	23	40%	4.5
7	26	48%	4.4	26	21%	2.7
8	29	31%	3.1	32	16%	2.6
9	41	35%	4.8	45	12%	2.7
10	100	0%	0	100	0%	0

* Tree assessed by one person.
** Tree assessed by two people filling out one assessment form.
Coeff. of Var. - Coefficient of Variation.
Con. Int. - Confidence Interval

maple can comprises 90% or more of the basal area, particularly in the central part of the province. The chips on the field template used to evaluate foliar colour were based on collections of sugar maple foliage. Therefore, there is a potential bias against other species. However, after three complete surveys the assessment crews have not expressed difficulty in applying the field templates to numerous deciduous tree species, although sugar maple was the target species and it made up just more than 75% of the trees encountered in the provincial survey. If the survey were to be adopted to target other tree species, for example, some of the Carolinian species in the extreme southwest or the birch and aspen forests of the northern boreal zone, then additional species-specific colour chips could be assembled so that the evaluator could select the appropriate template in the field. This has not, however, been recognized as a problem by the field staff for the current survey.

ONTARIO HARDWOOD FOREST HEALTH SURVEY RESULTS

In 1986 the Ontario Ministry of the Environment initiated a province-wide deciduous forest health survey using the DI method. The 1986 survey was used as the benchmark year.

The plots were re-assessed in 1987, 1989, and 1990. A survey was not conducted in 1988. Using a stratified systematic design, 110 permanent observation plots were established, each consisting of 100 trees greater than 10 cm dbh, for a survey total of 11,000 trees. Plot selection criteria ensured that the plots were as comparable as practical given the extent of the hardwood forest range in Ontario and the complex local and regional superficial geology. Only results current to 1989 will be discussed in this paper.

Three or four crews of two people each were used to assess the plots. The two person crews used the paired assessment approach. Tree assessment was conducted between the second week in July and the first week in September, starting at the most northerly plots and working south, thereby compensating for the north/south climate gradient in the province.

Quality assurance procedures included extensive crew training and overlap plots. The crews were trained for two weeks at a series of plots which contained a wide gradient of decline symptoms. They were then tested for consistency, before commencing the survey plot assessments, by a process similar to the original DI test trial. In addition, in 1989 seven plots were randomly selected as blind overlap plots. Four plots were assessed by two crews and three plots were assessed by three crews. In each case the crews were unaware of previous or subsequent assessment activity.

The results of the 1989 blind plot assessments are summarized in Table 3. Eight of the paired plot assessments had a mean DI which varied by 5 or less. Only three of the 13 paired plot assessments had a mean DI which varied by more than 10, the greatest being 15. Although the difference in mean DI at some plots was substantially higher than the original test trials would suggest should occur, none of the differences were statistically significant ($p > 0.05$). Experience has shown that the difference between tree comparisons can be affected by weather. Specifically, very bright sunny days can "backlight" the tree canopy which can sometimes give the foliage in the upper crown a slightly chlorotic hue. Similarly, uncomfortable working conditions and adverse weather may occasionally result in a somewhat hurried assessment with the resultant deterioration in accuracy and/or consistency.

One of the distinct advantages of the DI method is that it yields quantitative data, which makes the comparison of

Table 3. Results of replicate assessments of blind overlap plots during the 1989 Ontario forest health survey.

Plot Number	Crew A	Crew B/C	DI Difference	Sum of Squares	Mean Square Error	F Ratio
2	22	16	6	316	542	0.58
17	27	12	15	777	542	1.43
26	19	14	5	245	542	0.45
26	14	9	5	220	542	0.41
26	19	9	10	465	542	0.86
36	13	11	2	143	542	0.26
36	13	11	2	143	542	0.26
36	13	13	0	0.5	542	0.0009
57	9	8	1	52	542	0.10
84	9	7	2	55	542	1.10
107	25	12	13	634	542	1.17
107	25	12	13	638	542	1.18
107	25	25	0	4.5	542	0.0008

Plots 2, 17, 57 and 84 were overlapped by two crews, Plots 26, 36 and 107 by three crews. In no case was the difference in DI between crews statistically significant (p >0.05).

various data subsets more meaningful. Table 4 summarizes the DI for the 10 tree species that comprised at least 1 percent of the survey tally. Over the first four years of the survey, sugar maple and American beech averaged the lowest DI (best overall tree condition, DI=12), compared to white birch which averaged the highest DI (worst overall tree condition, DI=27). In the same time period black cherry varied the most, from a high of DI=30 in 1987 to a low of DI=15 in 1989. In the four year period from 1986 to 1989, six tree species (sugar and red maple, white ash, American beech, yellow birch and black cherry) showed a tendency to improve in overall condition, whereas three tree species (ironwood, red oak and white birch) showed a deteriorating trend. One species (basswood) deteriorated in 1987, but subsequently improved so that there was no absolute change in the DI in 1989 compared with the 1986 baseline year. These data have not been tested statistically, but since the change in DI over four years (Table 4) is comparable to the confidence interval of the DI for trees in that condition, it is unlikely that the apparent trends are significant. For example, red maple appeared to improve over the four years because the absolute DI changed in 1989 relative to 1986 (shown in Table 4 to be -4). However, from Table 2 the 99% confidence interval for a tree with a mean DI of 23, which is similar to the mean DI of 21 from red maple, is 4.5. Since the confidence interval is

Table 4. Mean decline index for tree species comprising at least 1% of the survey tally for the 1986, 1987, and 1989 Ontario forest health survey.[*]

Tree Species	% of Tally	Mean DI 1986	Mean DI 1987	Mean DI 1989	Mean DI 1986-1989	DI Change[**]
Sg Maple	75.0	12	14	10	12	-2
Wt Ash	3.5	17	18	13	16	-4
Rd Maple	3.1	22	24	18	21	-4
Am Beech	3.0	13	13	9	12	-4
Basswood	3.0	18	21	18	19	NC
Ironwood	2.6	23	22	31	25	+8
Yw Birch	1.6	20	24	18	21	-2
Rd Oak	1.4	20	17	24	20	+4
Bk Cherry	1.6	28	30	15	24	-13
Wt Birch	1.0	24	26	31	27	+7

[*] Based on a survey of 11,000 trees
[**] DI change in 1989 relative to 1986, a negative DI implies an improvement in tree condition.
NC - no change
Survey not conducted in 1988

similar to the overall change in DI, it is unlikely that the trend towards improvement is real. A few more years of survey data will be required before the "noise level" of the rating system can be quantified. It is also unlikely that an annual survey is required, or even desirable. Overall forest condition should change gradually. A two-to a five-year re-assessment frequency is likely adequate for tracking real change in forest health. The Ontario forest health survey was conducted annually (except for 1988) from 1986 to test the rating system for reproducibility and to assess its sensitivity in parts of the province where defoliating insect epidemics were expected.

Mortality of all surveyed trees averaged 1.1% in 1986, 3.3% in 1987 and 1.8% in 1989. The 1987 mortality figure is erroneously high, because some trees tallied as dead in 1987 were tallied as live in 1989. These trees were in plots which were located in areas where defoliation by forest tent caterpillar, bruce spanworm and/or gypsy moth had been severe in 1987 and 1988. Tree assessment was conducted in mid season, which should have been after refoliation had occurred on trees which had been defoliated in the spring or early summer. It is possible that some trees may not have had the vigour to refoliate by the time the survey was conducted and were mistakenly tallied as dead. Similarly, some trees devoid of foliage and near death as a result of defoliation stress, and so tallied as dead, may have responded the following season with epicormic sprouts and subsequently been tallied as live trees.

Hardwood forest covers an area of approximately 175,000 km^2 in Ontario. This expansive area incorporates several climate zones, at least nine forest sections as defined by Rowe (382), and more than two dozen Ontario government administrative districts. Table 5 summarizes three years of

Table 5. Mean decline index of the eight forest sections* in the survey area, based on results of the Ontario forest health surveys conducted in 1986, 1987, and 1988.

Forest Section	% of Survey Area	Mean DI 1986	Mean DI 1987	Mean DI 1989	Mean DI 1986-1989	DI Change**
Niagara	16	14	13	6	11	-8
Huron-Ontario	27	12	13	9	11	-3
Upper St. Lawrence	9	12	9	5	9	-7
Algonquin-Pontiac	8	15	23	15	18	NC
Middle Ottawa	10	13	14	10	12	-3
Georgian Bay	12	17	18	18	18	+1
Sudbury-North Bay	8	15	21	19	18	+4
Algoma	7	16	16	12	15	-4
Quetico	3	9	7	4	7	-5

* Forest Sections defined by Rowe, 1972.
** DI change in 1989 relative to 1986, a negative DI implies an improvement in tree condition.
NC - no change.
Survey not conducted in 1988.

survey data categorized by forest section. These sections are based primarily on species associations, which reflect gross climatic and soil factors. From 1986 to 1989, the Quetico forest section had the lowest mean DI (=7). By comparison, three forest sections, Algonquin-Pontiac, Georgian Bay, and Sudbury-North Bay, all had the highest mean DI (=18). These forest sections occupy the north central area of the deciduous forest in Ontario. They are Precambrian Shield sites, characterized by thin, coarse-textured or poorly drained soil. These are non-agricultural lands with many areas even classed as marginal for forest production. A higher DI, reflecting generally poorer forest condition, was expected for these regions. This is another indication that the DI system works, because it was able to identify expected regional differences in relative forest condition. The Niagara and the Algonquin-Pontiac sections were the most variable during the three year period; the former improved in tree condition with each survey year, and the latter deteriorated substantially in 1987 then subsequently improved in 1989. This was a trend observed at most plots, i.e., those that deteriorated in 1987 (generally) improved in 1989. Over the four year period, comparing 1989 to 1986, six forest sections improved in condition, two deteriorated, and one section, although it showed considerable annual fluctuation, did not change overall.

When the components that make up the DI were examined individually at the plots which deteriorated in 1987 it was found that, for the majority of the plots, the increase in DI was a result of an increase in the frequency of small and chlorotic foliage and not twig/branch dieback. This would imply a short-term stress from which tree recovery was possible. Recovery was evident in 1989. The plots where foliar abnormalities were most pronounced were generally from areas which had experienced defoliation. The foliar parameters reflect almost an immediate change in relative tree condition, and therefore, combined with assessment error, likely account for the majority of the annual fluctuation in individual plot mean DI. In contrast, twig/branch dieback, reflects a longer-term response and is indicative of a progressive deterioration in tree condition. Therefore, the DI system is not just a method for tracking tree deterioration, it is a method of assessing short- and long-term changes in tree condition and is sensitive enough to identify subtle differences in relative stand condition between regions.

For purposes of comparison with other surveys and to facilitate trend detection the DI gradient can be described in terms of five general decline categories:

Class	Decline Rating	DI Range
1	Healthy	<11
2	Trace	11-15.9
3	Low	16-20.9
4	Moderate	21-25
5	Severe	>25

The percentages of the survey plots in each of these decline categories for the first three survey years are illustrated in Figure 2. In 1989, 53% of the plots were in the healthy category. This compares with 21% in 1987 and 17% in 1986. In contrast, 6% were in the severe decline category in 1989, compared with 9% of the plots in 1987 and just less than 1% in 1986. Generally, there was a marginal deterioration in tree condition from 1986 to 1987 and a subsequent improvement in 1989. Over the period from 1986 to 1989, tree condition improved at more than twice as many plots as it deteriorated (the DI dropped one or more categories at 51% of the plots while the DI increased one or more categories at 20% of the plots). In theory, the DI system could be used to develop up to 11 decline classes, as the individual crown parameters which go into the DI scores are estimated to the nearest 10%. However, in practice, with the first three years of survey data, it is apparent that all of the plot mean DI scores are clustered in the lower half of the potential 1 to 100 range (mean plot DI<50), with most means falling below 25. It takes many very severely declining trees in a plot to skew the mean DI above 25, as all trees on the plot over 10 cm dbh are assessed and the younger trees are both more numerous and tend to be in better condition. Also, by keeping the number of relative decline classes small there is a greater likelihood of identifying real trends, instead of survey "noise". At this time, until further Ontario survey data are obtained, five decline classes compares favorably with other systems and is applicable for a preliminary examination of the trends.

Repeated assessment of the same trees with the DI methodology allows for quantitative spacial and temporal trend analysis. The recent introduction of a Geographic Information System for use with these data should enhance the effectiveness of the interpretation of the survey results. The mean DI for each plot for each year was assigned to one of the five previously described descriptive decline categories and mapped using the GIS. For each plot the GIS produces a Thiessen polygon, which is proportional to the survey area that the plot is intended to represent, and displays the polygons on a provincial map. Polygons are smaller in parts of the province where the plot density is high, and larger where there are fewer plots. Plot polygons can be shaded, using GIS, according to which of the five decline classes the plot mean DI represents. Figure 3 illustrates the distribution of the five

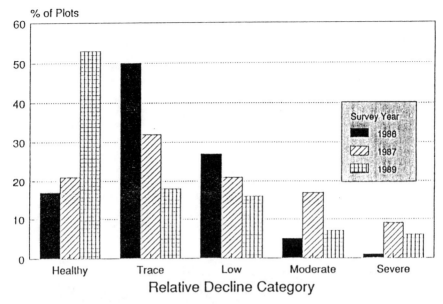

Fig. 2. Percentage of Ontario hardwood forest health survey plots in each of five relative descriptive decline classes for the 1986, 1987, and 1989 survey years. A survey was not conducted in 1988.

decline classes for 1986. Figure 4 illustrates the change in decline in 1989 relative to 1986. A decrease of one decline class means that the plots represented by a particular polygon have improved in condition when they were assessed in 1989 relative to the first survey in 1986. For example, a decrease of one decline class could mean the plot has moved from decline class 2, with a mean plot DI of between 11.0 and 15.9, to decline class 1, with a mean plot DI of less than 11. Conversely, an increase in decline classes would imply a deterioration in plot condition.

These figures clearly illustrate that most of the hardwood forest in Ontario is in the two lowest decline categories (Healthy and Trace decline). A regional pattern of forest condition is also evident. Forest condition is consistently good in the south-central part of the province along the Bruce Peninsula, the area north of Lake Ontario, and into the southeast. These are non-shield areas characterized by relatively deep, fine textured soil. Forest condition is consistently poorer in the central and north-central part of the hardwood forest range. These are Precambrian Shield sites

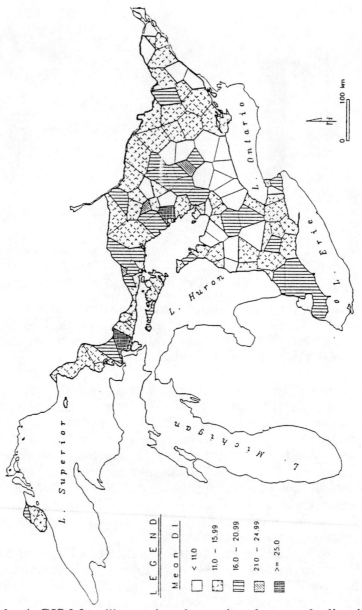

Fig. 3. A GIS Map illustrating the regional mean decline index (DI) based on the plots assessed during the 1986 Ontario hardwood forest health survey.

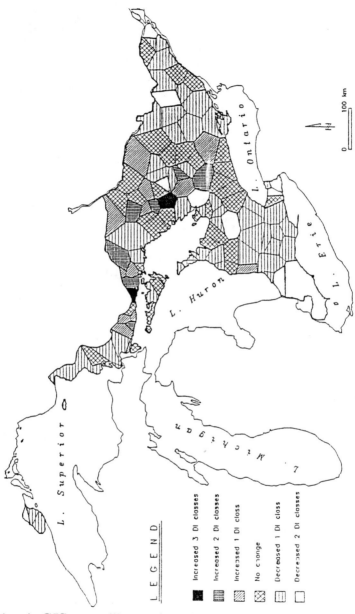

Fig. 4. A GIS map illustrating the change in regional mean decline index (DI) in 1989 compared with the 1986 baseline year.

where the soil depth and texture are much more variable. Sites where the soil is shallow and coarse, such as much of the Shield, are more limiting to forest growth because of restricted nutrient and moisture regimes, and therefore, (relatively) poorer tree conditions are to be expected.

CONCLUSIONS

The DI method of forest condition assessment has proven to be reliable and reproducible. It yields quantitative data to which descriptive statistics can be applied to enhance the interpretation of, and add confidence to, survey results. Subsequent surveys will enable forest managers to accurately quantify temporal and spacial trends in forest condition. A sound survey and monitoring system is integral in assessing the impact of atmospheric pollutants and other stresses on the forest ecosystem, and in evaluating the effect of legislated emission reductions.

PATTERN AND PROCESS OF MAPLE DIEBACK IN SOUTHERN Québec (CANADA)

G. Daoust[1,2], C. Ansseau[1,2], A. Thériault[1], and R. van Hulst[3]

[1]Dép. de biologie, Université de Sherbrooke
SHERBROOKE, QC, CANADA, J1K 2R1,

[2]Centre d'Applications et de Recherche en Télédétection (CARTEL), and

[3]Dept. of Biology, Bishop's University
LENNOXVILLE, QC, CANADA, J1M 1Z7

Dieback of forests in Europe and North America has been a growing concern, and, in spite of all the attention it has received, our understanding of the underlying mechanisms is still extremely inadequate (138, 267, 292, 314, 357). Dieback probably represents a complex of phytopathological phenomena with various single and multiple causes (256, 267, 357, 514). Therefore, traditional phytopathological approaches, oriented toward experimentation and analysis, may need to be supplemented by a more statistical and synthetic approach.

In Québec approximately 40% of the maple forests (the dominant deciduous forest in the province) are exhibiting some form of decline (136). Over half of Québec maple forests exhibit leaf loss of 11% or more (138).

The provincial government (Ministère de l'Energie et des Ressources, or MER) has documented the decline of maple forests over the last eight years using detailed studies of over 200 variously selected sites (267). Over the last five years essentially all maple forests in the province have also been surveyed once or twice by trained observers from a helicopter. The observers have noted the extent of leaf loss and discoloration, as well as any abnormal tree mortality, and these data have been used to construct several summary maps that indicate the extent of dieback (as estimated from these variables) and dieback-related mortality (70).

On these maps one can readily observe that the degree of damage varies non-randomly, although there are no obvious causative factors. In part, this is due to the nature of the maps produced. Obviously, maple forests do not cover 100% of the province, and mapping each individual maple forest on a general map was clearly not feasible. If areas with relatively low density of maple forests are left blank, the map properly locates the position and general state of the forest, but inadequately portrays the overall patterns in dieback.

Just as the careful study of the geographical variation of the incidence of human disease has provided both a powerful heuristic tool in epidemiology and a guideline for public policy development (7), the study of the spatial structure of forest dieback may be expected to provide a useful guide in the search for the causes of forest decline and in regulatory action. We have used the most complete data available to date on dieback in Québec maple forests: the province-wide maps of dieback based on the 1985/1986 reconnaissance flights (70) and the maps originally used by the MER for entering dieback data. Our aims were, first, to study the spatial structure of the dieback phenomenon, and to prepare a map of interpolated values of the extent of dieback to be expected in areas where there are presently no maple forests, but where isolated sugar maples and red maples often do occur. Our second aim was to use this map and the theory on which it is based in further research on the factors responsible for the dieback phenomenon and for planning purposes. In particular, we were interested in comparing the incidence of maple dieback at various locations to site quality.

Detailed maps of forest growth site classes were available from Lands Directorate, Environment Canada. These classes are based on increasing limitations to commercial wood production due to unfavorable soil, drainage, and climatic factors, with the highest scores indicating the greatest number of impediments to tree growth. If maple dieback is more severe in poor soils of low productivity, as one would expect, this should be shown clearly in a comparison of the spatial structures of dieback and site limitations. Note that the spatial autocorrelation in both variables prevents us from studying the correlation between them directly (276).

Other such comparisons may be of interest as well, but a statistically appropriate analysis of the relationship between any two variables that are spatially related has to employ the techniques of geostatistics. Such studies may lead to a more complete knowledge of the spatial distribution of forest dieback, and they will also facilitate follow-up studies using remote sensing (105) or other techniques (119).

METHODS AND RESULTS

We digitized the map of maple dieback in southern Québec prepared on the basis of the 1985 and 1986 surveys (scale 1:500,000) on a 10 x 10 mm grid, using a combination of the original scales for extent of leaf loss (1 = 0-10%, 2 = 11-25%, 3 = 26-50%, 4 = 51-75%, 5 = 76-100%) and mortality (little or no mortality vs significant mortality) to yield a 7-point scale, as follows: 1 = leaf loss < 10%, no mortality; 2 = leaf loss between 10 and 25%, no mortality; 3 = as before, but with significant mortality; 4 = leaf loss between 25 and 50%, no mortality; 5 = as before, with mortality; 6 = leaf loss greater than 50%.

Maps for the same region for *Land capability for forestry* were similarly digitized to obtain data on site quality. The index used is the map index, which assesses the severity of the constraints on forest tree growth. Note therefore that the index is really one of *site constraints:* the lower the index, the fewer constraints there should be on tree growth. The data thus gathered were analyzed using the GEO-EAS and GS+ software packages (119, 139), to derive information on the spatial structure of dieback and of site quality.

The appropriate method of interpolation for spatially distributed data is kriging on the basis of variograms (232). A variogram is a function that relates variation in a factor of interest (here, dieback intensity or site quality) and increasing distance between two sampling points, and this in different directions. In general, a variogram can be characterized by three parameters: the range, the sill and the nugget. The dependent variable, which measures the degree of variation, is known as the semi-variance, and the function is known as a semi-variogram or, simply, variogram. The semi-variance generally increases with distance, reflecting the fact that two increasingly distant points become less and less alike. At a certain distance the semi-variance often stabilizes; this distance is know as the range, and it indicates the scale of the spatial structure in the direction chosen, in the sense that points closer together than the range are spatially dependent. The level of the semi-variance that is reached is known as the sill. If a sill is present in a variogram this indicates that observations that are separated by a distance greater than the range can be treated as being spatially independent. The Y-intercept of the variogram is known as the nugget. A nugget greater than zero simply implies that there exists in the data at the sampling scale used some residual random variation that is spatially independent. For a more detailed discussion of these matters the reader is referred to references 103, 224, 232, 276, and 477.

The estimated variograms were subsequently used to construct interpolated values for the extent of dieback and of site quality using kriging (a spatial interpolation technique appropriate in the presence of spatial autocorrelation; see 276, 103). To verify the values thus obtained we used cross-validation (119, 224, 232).

The results of the interpolations were used for plotting contour maps of dieback, and for further calculations. The program used for contouring was TECPLOT (20). The resulting map of the dieback distribution in southern Québec is reproduced as Figure 1. Because of the absence of data for some areas, we were forced to use polygons delimiting areas of interest. This is the reason that some areas in Figure 1, notably the St. Lawrence Lowlands and Montréal, have not been treated. Proportions of the total study area in each of five of the classes previously established or devoid of maple forests are indicated in Table 1, as are the corresponding proportions for the data predicted using kriging. Note that proportions for the two data sets are not expected to be identical. This would only be the case if the areas where there is presently no maple forest would be allocated to the different dieback classes proportionally to the overall relative abundance of these classes. This may not be the case. In areas of high dieback, the empty regions on the map would likely be assigned to high dieback classes.

The variograms found to best summarize the spatial

structure of the dieback and site quality are listed in Table 2. Figure 2 illustrates variograms b and c (see Table 2) for the southeastern part of the region displayed in Figure 1. Figures 1, 3, and 4 represent the kriged maps obtained using the variogram parameters of Table 2 a, b, and c, respectively. The dieback data exhibit a distinct spatial structure, and the variogram fits the data rather well. The site quality data, on the other hand, indicate very little spatial structure. The map of interpolated values for the index is nevertheless the best that can be obtained from these data.

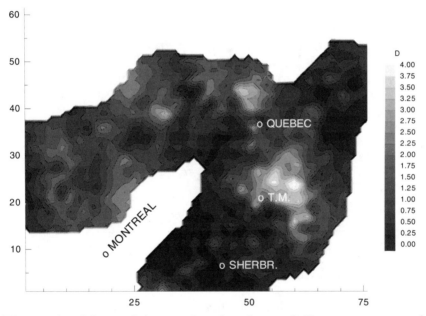

Figure 1. Map of interpolated values of D, a measure of maple decline, in southern Québec. High values of D (and light colors) indicate a high degree of dieback. The map is based on the 1985/1986 map of the Québec Ministère de l'Energie et des Ressources, and a composite measure of maple decline and mortality D, as discussed in the text. The interpolation technique used was block kriging on the basis of the variogram listed in Table 2a. The coordinates are synthetic, and bear no relationship to geographical coordinates. The geographical coordinates of the point (1,1) are 45 00'N, 74 45'W. The sampling interval is 5 km.

Table 1. Proportion of the total area of southern Québec in each of the dieback classes based on digitized 1985/1986 data and interpolated kriged data.

source	Dieback Classes				
	1	2	3	4	5
1985/86 data	0.47	0.33	0.15	0.02	0.02
kriged data	0.32	0.58	0.10	0.01	0.00

Table 2. Variograms used for preparing the interpolated dieback data of Figures 1, 3, and 4. The variograms that correspond to b and c are illustrated in Figure 2. One map unit corresponds to 5 km.

variogram	model	nugget	sill	range
a (D, Fig. 1)	spherical	0.25	0.17	7.0
b (D, Fig. 2)	spherical	0.27	0.39	11.95
c (F, Fig. 2)	linear	0.18	.	.

DISCUSSION AND CONCLUSIONS

There appears to be an inverse relationship between dieback and site quality. The poorer sites are situated mostly north of the St. Lawrence valley in the Laurentide Hills, at higher elevations in the southern and southeastern part of the province, and in a band of serpentine soils near Thetford-Mines. These correspond to the lighter areas in Figure 1, the areas with the greatest amount of dieback. In the Laurentide Hills, soils are predominantly podzols, derived from acidic Canadian Shield rocks (462). At higher elevations soils are often shallow and poor due to leaching (462) and serpentine soils are notably infertile or even toxic (505). It is clear from our analysis that the spatial distribution of dieback is far from

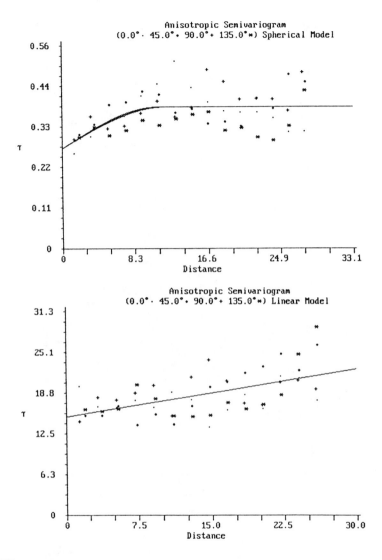

Figure 2. Variograms for the southeastern part of the region displayed in Figure 1. The upper variogram is for the dieback index (1985 data) for this subregion alone. The variogram that best fitted the site index data for the same region is shown in the lower variogram.

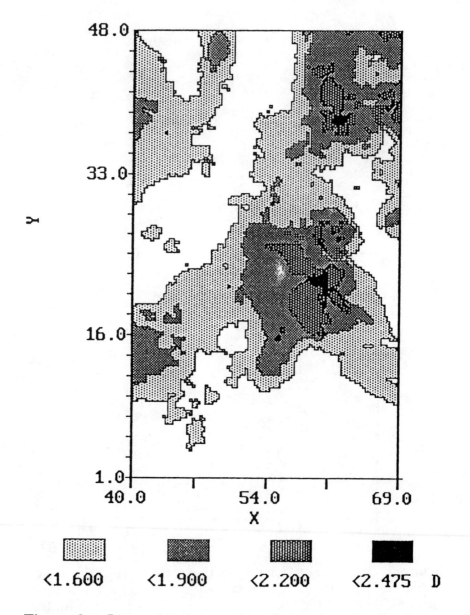

Figure 3. Interpolated map based on the dieback index variogram for the southeastern part of the region displayed in Figure 1. The sampling interval is 5 km.

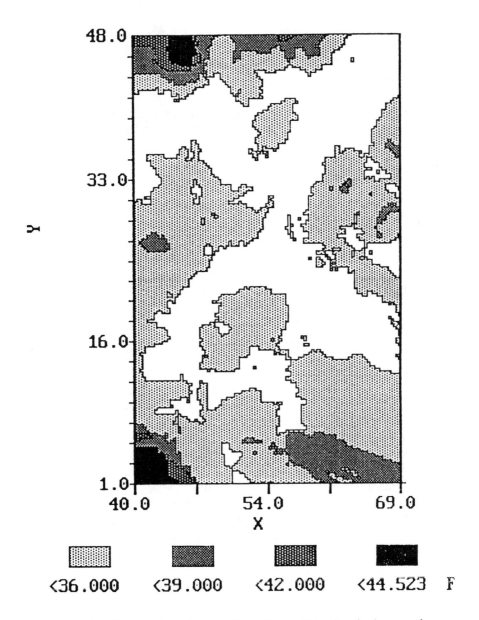

Figure 4. Interpolated map based on the site index variogrm for the southeastern part of the region displayed in Figure 1. The sampling interval is 5 km.

random, and that forests that are up to 35 km (7 units of 5 km) from one another tend to show a similar extent of dieback (the range of variogram (**a**) in Table 2).

While there exists, therefore, a clear spatial structure in the extent of maple dieback, the reasons for this are not obvious. Evidently, numerous environmental factors exhibit a spatial structure (e.g. 478), but so does, in all likelihood, the genetic structure of maple populations. In fact, evidence for spatial variation in seed protein traits has already been found in the case of the red maple (461).

Figures 2, 3, and 4 are the variograms and the corresponding interpolated maps for part of the study region of maple dieback and site quality, respectively. This part of the study region exhibits both very high and very low incidences of maple dieback (see Figure 1). The variogram fitted for the dieback data in this restricted data set and the resulting interpolation map are somewhat different from the variogram fitted for the full data set and the corresponding map (cf. Table 1 (a) and (b), and Figures 1 and 3). These differences are a result of the fact that the data used are quite different: Figure 1 is based on data from all of southern Québec, while Figure 3 uses only data from the smaller Thetford Mines subregion. Consequently, spatial structures that are much more extensive could be incorporated in the former, as compared to the latter, and the detailed patterns in the resulting kriged maps are different, although the major patterns resemble one another. Since site quality data were available for the subregion only, we were forced to reanalyze the dieback data at this scale as well.

The site index data and the dieback data present quite different pictures. Examination of the best variogram that could be fitted to the site data points (see Figure 2) shows that the site quality data, unlike the dieback data, do not exhibit an unambiguous spatial structure. The fitted variogram is linear with a very small slope, which means that, at this scale, a spatial structure is virtually non-existent. We conclude that, in this part of our study region at least, maple dieback is quite unrelated to site quality, as defined by our index.

If one examines a posting of the site index data for the restricted area, the index shows little variability (Figure 5). The central part of the subregion studied has a somewhat lower index value than the area on its periphery. Yet, it is in the center that maple dieback is at its highest! The solution

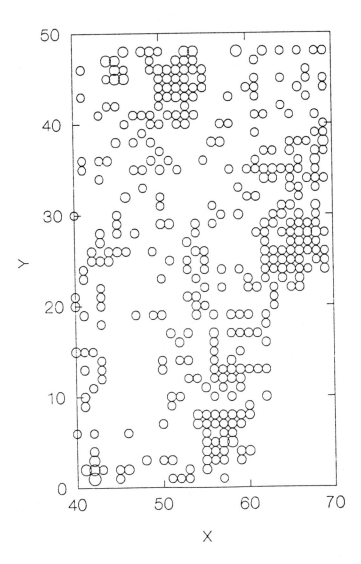

Figure 5. A posting of the site quality data for the subregion studied. The diameter of each circle indicates the site index for the site sampled.

to this paradox may be that the site index is not based on actual tree growth, but on the extent of tree growth predicted on the basis of climatic and topographic variables, elevation, soil nutrients, thickness of the humus layer, etc. The region in the center of our study area is characterized by serpentine soils, however, and the actual tree growth in these infertile soils is almost certainly overestimated by the site index. Another possible explanation for our findings is the demonstrated positive relationship between less acidic soil with a high quality humus and increasing phosphorus deficiencies, which, in turn, are related to progressive decline in sugar maple stands (347). Further work is clearly in order here.

The inadequacies of the site quality index used should also caution us to view the lack of agreement between the spatial structure of maple dieback and that of the site index with some circumspection. The example does show, however, that the geostatistical methods used here can be powerful tools in disentangling the complex causal relationships that characterize the dieback phenomenon. A more detailed analysis relating maple dieback to several environmental factors using geographical information systems and geostatistical methods, has the potential to be much more instructive than this explorative study using one synthetic variable.

The dieback data used can also be criticized on the basis of their unclear and subjective relationship to the original data, collected visually using aerial reconnaissance, and entered on maps at the scale of 1:50,000. To investigate the degree of distortion introduced by the transcription process we have also digitized the original maps, using point sampling, with interpoint distances of 2.5 km. The data thus obtained were compared to the data obtained from the synthetic map, and both the average proportions of the maple forests in the dieback categories and the models fitted for the variograms, were essentially identical.

The use of spatial statistics to analyze dieback data offers a double reward. First, the maps that can be produced using kriging may be easier to interpret and use than maps indicating only the extent of dieback at sampling stations. Spatial gradients in dieback extent that were not visible in the original map appear clearly in the kriged map (Figure 1). Second, once the null hypothesis of a random spatial structure has been rejected (as it was in the case of the dieback data), we can ask the more interesting question of what other environmental patterns correlate with the interpolated data. We have shown here how such an analysis begins by comparing the spatial structure in each variable. If an environmental variable does demonstrate a clear spatial structure, the relationship between the two variables can be explored further by co-kriging (232, 477). This may also lead to a more accurate spatial prediction of the key variable in question, the dieback index. Geostatistical methods are powerful tools that may shed a new light on the multiple causes of forest dieback.

ACKNOWLEDGEMENTS

We thank the FCAR (scholarship to G.D.), the Ministère de l'Energie et des Ressources (grant to C.A.), and NSERC (grant to RvH) for financial assistance.

DEVELOPMENT OF A HAZARD RATING SYSTEM FOR DECLINE IN SOUTHERN BOTTOMLAND OAKS

Vernon Ammon[1], T. Evan Nebeker[1], Carolyn R. Boyle[1], Francis I. McCracken[2] and James D. Solomon[2]

[1]Professor, Professor, and Senior Research Assistant, respectively, Mississippi Agricultural and Forestry Experiment Station, Mississippi State University, Mississippi State, MS and [2]Research Pathologist and Principal Research Entomologist, respectively, USDA Forest Service, Stoneville, MS

Oak forests are among the most important timber resource in the United States, providing recreational opportunities to millions, habitat to thousands of species of flora and fauna, and a third of the nation's hardwood sawtimber volume. Oak forests occupy about 38 percent of the forest land in the eastern states. A large block of mixed forest types extends along the Appalachian Mountain range from New England to Mississippi and Alabama. Another block extends from Minnesota to Texas. Oak-pine forest types are most common in southern areas, particularly Tennessee, Georgia, Alabama, and Mississippi. About 20 oak species are considered commercially important, although five species account for most of the volume harvested.

Stresses impacting this resource eventually generate losses unacceptable to various private and public user groups. Even though decline and death of oaks is a natural, inescapable, and orderly event, public concern intensifies when it occurs in ways that fail to meet their expectations for a given species on a particular site. The history of decline and death of oaks in the

United States has been documented in reports covering more than 130 years. From 1856 to 1981, more than 26 decline events were reported from eight eastern states affecting almost all species of oaks. Fourteen factors have been implicated as either primary or secondary agents responsible for decline and mortality. These include late freeze (50), drought (36, 50, 280, 309, 442, 453, 467) twolined chestnut borers (92, 117, 442), gypsy moth (36, 82, 434), Armillaria root rot (36, 125, 280), Hypoxylon canker (277), Ganoderma root rot (280), aspect (147, 245), slope (147, 468), coarse, shallow, poor soils (221, 442), encroachment of civilization (259), and changing climate (187).

Oak decline is still a disease of unknown etiology. The disease is apparently caused by a number of abiotic and biotic agents in a manner not fully understood. Research to describe conditions associated with declining trees is necessary to reduce losses to levels which satisfy resource management objectives. Research results to date have not provided the information required to satisfactorily deal with oak decline. Failure to fully understand the factors responsible for this phenomenon and the interactive effects of host, stand, site, climate, and pests are of considerable concern. The ability to predict stand susceptibility and potential losses requires an expanded data base. Research challenges include the need to identify factors involved with decline, to identify where and how these factors contribute to decline, and to evaluate their impact, individually and collectively.

The general hypothesis that we put forward is that oak decline in southern hardwood bottomlands is associated with specific site and stand characteristics which can be measured and utilized in analytical techniques to separate decline from non-decline (healthy) sites. The first objective was to collect biological and edaphic data from decline and non-decline plots. A decline plot is defined as an area with at least one southern red oak exhibiting characteristic decline symptoms; those symptoms include terminal death (dieback) and defoliation. The second objective was to subject these data to a canonical discriminant analysis (255) in an attempt to separate decline from non-decline (healthy) plots. The third objective was to develop a model (hazard rating) that would identify plots with characteristics similar to those we established. These sites might warrant early silvicultural treatment to reduce the impact from decline in a particular area.

MATERIALS AND METHODS

The midsouth hardwood forest was stratified along two main drainage systems; the Mississippi and the Tennessee-Tombigbee Rivers. Within each drainage system, variable radius plots containing declining (defined above) and non-declining trees were established using a 10 basal area factor angle gauge. Tree and site data were recorded for each plot (Tables 1 and 2). Data were collected from over 3,000 trees growing in 272 field plots at 22 locations in 7 states (Figure 1). One hundred sixty-eight field plots contained trees in various stages of decline and 104 contained trees showing no evidence of decline. One hundred fifty one plots were established at 10 locations in the Mississippi River drainage basin and 121 plots at 12 locations were established in the Tennessee-Tombigbee River drainage basin.

Statistical Analyses

Statistical analyses were performed using unweighted data. Data were treated to independent sample t-tests and chi-square tests of independence to determine if decline and non-decline plots had similar tree and site characteristics.

Canonical discriminant analysis (255) was used to determine the subset of these variables that best differentiated between decline and non-decline plots. Starting with those variables judged to have the greatest potential for discriminating between the two types of plots, a stepwise selection procedure was used to find the "best" variables for inclusion in the discriminant function. These calculations were performed with the SAS STEPDISC procedure (18) using the Wilkes method and default tolerance values. The SAS DISCRIM procedure with the CANONICAL option was then used to compute the canonical discriminant functions.

RESULTS AND DISCUSSION

Tennessee-Tombigbee River Basin

Results of the t-tests were similar for most variables for decline and non-decline plots (Table 3). Those plots identified as decline sites contained trees with significantly fewer sticks, had significantly fewer healthy trees (crown condition class 1),

Table 1. Tree variables measured on each decline and non-decline plot.

Variable	Recorded
Species	Species (common name)
Diameter	Inches at breast height (DBH)
Form	1 = Straight tree, no major branching 2 = Some stem curvature and/or major branching in first log 3 = Crooked or hollow tree
Grade	1 = Logs that yield 60% 1 common and better lumber 2 = Logs that yield 40-60% 1 common and better grade lumber 3 = Logs that yield 20-40% 1 common and better grade lumber 4 = Logs that yield < 20% 1 common and better lumber
Logs	Number of 16 ft logs or trees over 10 in DBH
Sticks	Number of sticks (5 ft cylinder of wood at least 4 in diameter for trees under 10 in DBH)
Crown Class	1 = Dominant 2 = Co-Dominant 3 = Intermediate 4 = Suppressed
Crown Condition	1 = Healty tree 2 = <1/3 of crown affected 3 = 1/3-2/3 of crown affected 4 = >2/3 of crown affected 5 = Dead tree
Growth	Millimeters (last year, 5 years, 10 years)

Table 2. Site variables measured on each decline and non-decline plot.

Variable	Recorded
Slope	Percent
Aspect	0 = NE 1 = NE 2 = E 3 = SE 4 = S 5 = SW 6 = W 7 = NW
Topographic Position	1 = Upland ridge 2 = Upland sideslope 3 = Upland bench 4 = Upland bottom 5 = Bottomland slough 6 = Bottomland flat 7 = Bottomland ridge 8 = Bottomland terrace
Basal Area	Square feet
Site Index	Height in ft and age 50 yr of three dominant/co-dominant trees
Soil Components	pH, percent organic matter, cation exchange capacity
Mineral Analysis	N, P, K, Ca, Mg, S, and Zn

deteriorating crown conditions and had higher basal areas than non-decline plots. Chi-squared analysis demonstrated no significant difference in slope, aspect or topographic between decline and non-decline plots.

The variables used in the stepwise selection procedure are listed in Table 4. The same variables were used for both the Tennessee-Tombigbee River basin and the Mississippi River basin. The canonical correlation is equivalent to the multiple regression correlation of regression analysis and, when squared, gives the total percent of variation in plot differences explained by this group of variables. Stepwise discriminant analysis selected seven variables that explained 85.40% of the variation between decline and non-decline plots in the Tennessee-Tombigbee River basin. The variable which contributed most to differentiating between plots was percent crown condition 1 (69.7%). An additional 7.27% contribution was provided by tree growth during the past five years. The tree variables percent crown class 3, percent form 2, percent crown class 2, and crown condition 5 each explained approximately 2% of the variation between decline and non-decline plots.

The variables selected for creating the canonical discriminant function describing plots within the Tennessee-Tombigbee River basin are in Table 5. The canonical coefficients are linear combinations of discriminating variables that relate the individual variables to the function. The magnitude of each coefficient gives the variable's relative contribution to calculating the discriminant score. The actual signs of the coefficients are arbitrary, that is they depend on the signs of the other coefficients. Comparison of the centroids shows that decline plots are distinctly separated from non-decline plots with a minus value as opposed to a positive value. By comparing the sign of the canonical coefficients to that of the centroids, we can conclude that decline plots in the Tennessee-Tombigbee River basin are characterized by a tendency for growth for the last five years to be higher, a decrease in the number of trees with some stem curvature and/or major branching in the first log, a decrease in the number of co-dominant and intermediate trees, a decrease in the number of healthy trees, and increases in the number of dead trees and trees with over two-thirds of their crowns in a deteriorating condition.

The best model for the Tennessee-Tombigbee River basin data correctly classified 90.6% of the decline plots and 100.0% of the non-decline plots for which all the discriminating variables were recorded.

Mississippi River Basin

T-test statistics were similar for most variables (Table 6). These same tests detected twelve variables which were significantly different ($P<0.05$ or <0.01) when data from decline and non-decline plots were compared. A significantly ($P<0.05$) larger number of red oaks were growing on decline

Table 3. Variable means for decline and non-decline plots located in the Tennessee-Tombigbee River basin.

Variable	Decline (63 plots)	Non-Decline (58 plots)	t-test p-value
White oaks(%)	18.9	22.8	ns
Red oaks(%)	44.3	38.7	ns
Other species(%)	36.8	38.3	ns
Tree diameter(in)	13.9	13.6	ns
Form 1(%)	53.3	60.9	ns
Form 2(%)	19.1	17.9	ns
Form 3(%)	9.9	5.5	ns
Logs (number)	1.0	1.0	ns
Sticks (number)	3.5	4.2	0.05
Crown class 1(%)	17.0	14.9	ns
Crown class 2(%)	45.1	47.1	ns
Crown class 3(%)	15.3	18.3	ns
Crown class 4(%)	22.3	19.4	ns
Crown condition 1(%)	63.7	99.0	0.01
Crown condition 2(%)	16.3	0.9	0.01
Crown condition 3(%)	8.9	0.0	0.01
Crown condition 4(%)	2.9	0.0	0.01
Crown condition 5(%)	8.1	0.0	0.01
Growth last yr (mm)	1.2	1.1	ns
Growth last 5 yr (mm)	4.7	4.3	ns
Growth last 10 yr (mm)	9.8	9.0	ns
SITE VARIABLES			
Basal area (sq. ft)	96.9	85.9	0.05
Site index (50 yr, ft)	76.7	80.0	ns
Organic matter (%)	1.5	1.5	ns
pH	4.9	4.9	ns
Cation Ex Capacity (meq)	8.9	10.3	ns

Table 4. Variables introduced for computer selection

Variable

Slope (%)
Aspect[1]
Topographic position[1]
Basal area (sq. ft.)
Site index[1]
Growth last year (mm)
Growth last 5 years (mm)
Growth last 10 years (mm)
Organic matter (%)
pH
Cation exchange capacity (meq/100g)
N, P. K, Ca, Mg, S, and Zn (lbs/ac)
Trees (number)
Diameter breast height (inches)
Logs (number)
Sticks (number)
White oaks (%)
Red oaks (%)
Other trees (%)
Form 1 (%)
Form 2 (%)
Form 3 (%)
Grade 1 (%)
Grade 2 (%)
Grade 3 (%)
Grade 4 (%)
Crown Class 1 (%)
Crown Class 2 (%)
Crown Class 3 (%)
Crown Class 4 (%)
Crown Condition 1 (%)
Crown Condition 2 (%)
Crown Condition 3 (%)
Crown Condition 4 (%)
Crown Condition 5 (%)

---

[1] See Tables 1 and 2 for units of measurement.

Table 5. Canonical discriminant function coefficient statistics, Tennesee-Tombigbee River basin.

Variable	Standardized Canonical Coefficient
Growth last 5 years	-0.61915
Form 2 (%)	0.29738
Crown class 2 (%)	0.25154
Crown class 3 (%)	0.42033
Crown condition 1 (%)	0.72679
Crown condition 4 (%)	-0.18468
Crown condition 5 (%)	-0.08096

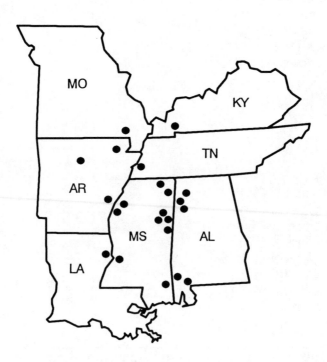

Figure 1. Distribution of oak decline plots in southern United States.

Table 6. Variable means for decline and non-decline plots located in the Mississippi River basin.

Variable	Variable Means		t-test p-value
	Decline (104 plots)	Non-Decline (46 plots)	
White oaks (%)	10.4	11.6	ns
Red Oaks (%)	61.5	53.5	0.05
Other species (%)	28.0	34.9	ns
Tree diameter (in)	18.6	19.0	ns
Form 1 (%)	58.7	63.0	ns
Form 2 (%)	12.1	15.1	ns
Form 3 (%)	15.0	15.6	ns
Logs (number)	1.6	1.6	ns
Sticks (number)	1.0	1.3	ns
Crown class 1 (%)	26.3	28.9	ns
Crown class 2 (%)	40.8	38.0	ns
Crown class 3 (%)	20.8	20.0	ns
Crown class 4 (%)	11.7	13.0	ns
Crown condition 1 (%)	27.9	97.5	0.01
Crown condition 2 (%)	51.3	1.9	0.01
Crown condition 3 (%)	8.7	0.6	0.01
Crown condition 4 (%)	1.2	0.0	0.01
Crown condition 5 (%)	10.7	0.0	0.01
Growth last yr (mm)	1.5	2.5	0.01
Growth last 5 yr (mm)	8.1	12.8	0.01
Growth last 10 yr (mm)	16.5	26.5	0.01
SITE VARIABLES			
Basal area (sq. ft)	121.7	116.9	ns
Site index (50 yr, ft)	96.0	93.6	ns
Organic matter (%)	1.7	2.1	0.01
pH	5.5	5.6	0.05
Cation Ex Capacity (meq)	21.1	21.2	ns
Sulfur (lbs/ac)	246.9	297.4	0.01

plots. Decline plots also contained significantly higher ($P<0.01$) number of trees with dieback and defoliation (crown conditions 2, 3, and 4) and dead trees (crown condition 5). Growth of trees on decline plots for the last year, last five years, and last ten years was significantly ($P<0.01$) less. Soil on decline plots contained less organic matter, the pH was lower, and contained more sulfur. Slope aspect and topographic variables were not significantly different between decline and non-decline plots.

Discriminant analysis procedures were again used to examine differences between decline and non-decline plots with respect to several variables simultaneously. The variable list used appears in Table 4. Stepwise discriminant analysis identified ten variables that explained 82.48% of the variation between decline and non-decline plots in the Mississippi River basin. Percent crown condition 1 contributed over 74% to this differentiation. Minor contributions (<1% each) were made by seven variables including percent grade 2, pH, percent dominant trees, trees with less than 1/3 of their crowns affected, trees with over 2/3 of their crowns with dieback and defoliation, grade 1, and soil sulfur content. Percent grade 4 trees and growth during the last year were variables that contributed from 2% to over 3% respectively to differentiating decline from non-decline plots in the Mississippi River basin.

The variables selected for developing the canonical discriminant function for the Mississippi River basin data are in Table 7. Again, by comparing the sign of the canonical coefficients to that of the centroid of the decline plots, we can conclude that decline plots in the Mississippi River basin are characterized by a tendency for soil sulfur and pH to be higher; for last year's growth, number of dominant trees, and number of healthy trees to be lower; for percentage of trees with dieback and defoliation affecting from less than 1/3 to over 2/3 of their crowns to increase; and for the percentage of 1, 2, and 4 grade trees to be lower.

The best model for the Mississippi River basin data correctly classified 99.5% of the decline plots and 100.0% of the non-decline plots for which all of the discriminating variables were recorded.

SUMMARY AND DISCUSSION

One approach to better understanding and subsequently managing oak decline, a disease of unknown etiology, is to

Table 7. Canonical discriminant function coefficient statistics, Mississippi River basin

Variable	Standardized Canonical Coefficient
Sulfur	-0.23006
pH	-0.33346
Growth last year	0.68098
Crown class 1 (%)	0.26755
Crown condition 1 (%)	1.50415
Crown condition 2 (%)	-0.68969
Crown condition 4 (%)	-0.20939
Grade 1 (%)	0.24166
Grade 2 (%)	0.36342
Grade 4 (%)	0.18835

identify and measure specific site and stand characteristics associated with decline sites. Discriminant procedures are mathematical, computer-assisted techniques which allow this approach to be attempted and assist in the development of a model which can be used in classifying sites as decline or non-decline. The data to test our hypothesis that site and stand statistics from decline sites differ significantly from non-decline sites was collected from over 3,000 trees growing in 272 field plots at 22 locations in 7 states in the Tennessee-Tombigbee River drainage basin and the Mississippi River drainage basin. Simultaneous examination of variables selected as important in differentiating decline plots from non-decline plots provided a set of seven variables for the Tennessee-Tombigbee River basin data that explained approximately 85% of the differences between decline and non-decline plots. Similar analysis of the Mississippi River basin data identified 10 variables that accounted for approximately 82% of the variation between decline and non-decline plots. Three variables (crown condition, growth, and class) were identified as important in differentiating decline from non-decline plots in both river systems. The importance of crown condition in the model is not unexpected since the plots were initially classified based on crown condition. Four additional variables were identified but not shared by both data sets. The form variable was part of the model for the Tennessee-Tombigbee basin but not the Mississippi model; whereas the variables grade, soil sulphur content, and soil pH were part of the Mississippi model but not the Tennessee-Tombigbee model.

The information presented here: i) establishes a baseline to which decline models generated elsewhere can be compared, ii) provides preliminary information that will help foresters and pest management specialists begin to identify stand and site variables most likely to characterize sites as decline or non-decline, and iii) demonstrates a computer-assisted approach to understanding this extremely complex disease. Inclusion of additional biological, environmental, and edaphic data is needed to strengthen the existing data base and analyses. The ultimate aim is to produce silvicultural guidelines to reduce losses to oak decline.

FOREST DECLINE CONCEPTS: AN OVERVIEW

Paul D. Manion and Denis Lachance

State University of New York, College of Environmental Science and Forestry, Syracuse, NY 13210 and Forestry Canada, Laurentian Forestry Centre, Sainte Foy, Quebec, Canada G1V 4C7

Forest decline concepts are not founded on a universal set of standards. There are many differences of opinion and some outright rejection of the concepts as too nebulous or confusing to be of any value. The populations of trees of concern have minimal management or are natural populations as contrasted to highly managed agricultural crops. There is potentially a need for a modified disease concept for this complex population. There is also a growing environmental awareness by the public that is expressed as a general acceptance that air pollutants are causing forest decline.

FOREST DECLINE, THE GERM THEORY, AND ENVIRONMENTAL POLLUTION

A forest decline concept, the subject of this book, does not necessarily displace the germ theory nor should it pool an array of problems identified by generic symptoms that are best understood as distinct problems. Kandler (chapter 4), for example, points out some of the fallacies of interpreting German forest problems as declines. Classic "Waldsterben" as perceived by the general public does not exist.

A forest decline concept should address problems that cannot be properly interpreted with a single agent (cause leading to an effect) germ theory. Although declines are characterized by a gradual, general loss of vigor and eventual death of a forest or of a tree species, single agent cause/effect diseases such as Dutch elm disease, chestnut blight and all other lethal diseases are likewise characterized by these same features. A decline concept is not needed for these. Decline concepts are needed for problems associated with an array of interacting biotic and abiotic factors (295). Muller-Dombois (chapter 2) and Manion (295) would suggest that a decline concept would also have to include a maturity, senescence, or aging component of the affected population.

A forest decline concept should likewise not be confused with the media-popularized forest "death" issues of the environmental pollution movement. Modern "wise men" extol "wrath of man" as the explanation for tree problems for "modern primitive" people.

In this connection, note the *a priori* recognition of air pollutants as a cause of forest decline in the introduction to the chapter by McLaughlin et al. (chapter 8). But, consider also the key points and evidence for air pollution involvement in the chapters by Kandler (chapter 4) and Skelly (chapter 5). Is there a genuine involvement of air pollution in forest decline or a paranoia that is inadvertently strengthened and perpetuated through introductory comments by well-intentioned scientists?

Conversely, one might consider the rational argument provided by Peterman (354) when he suggests that we should consider the ecological cost of making the judgement that there is no affect of air pollution on forest decline. Our statistical analysis of data is usually characterized by an attempt to minimize the Type I error by rejecting the null hypothesis at the 0.05 level. Rejection of the null hypothesis is usually interpreted as the statistical demonstration of an effect. One should recognize that acceptance of the null hypothesis (there is no effect) also has an error. This Type II error is inversely related to the type I error. Few acknowledge the magnitude of the Type II errors when they suggest that the evidence does not demonstrate an effect of air pollutants on forest decline.

Of course one has to be careful with this type of argument. Statistical analysis of an irrational hypothesis would undoubtedly lead to the acceptance of the null hypothesis with some probability of error and therefore some credibility for the nonsense hypothesis. This points out the dilemma encountered by Skelly (chapter 5). Rational science has not demonstrated the involvement of air pollutants in forest decline. The key examples are essentially misrepresentations of other problems. On the other hand, Skelly, as an air pollution specialist, recognizes that air pollutants can affect plants. Do we proceed as Woodwell (515) suggests and abandon our scientific principles to rely on "experience" for a higher purpose?

FOREST DECLINE TERMS, KEY FACTORS, AND MODELS

The concepts of forest decline have had only a few decades to develop and evolve. The contributors to the concepts clearly have different terms, limitations, key factors, models, and applications for their interpretations. You should recognize that decline precedes dieback in the Muller-Dombois (chapter 2) interpretation. Houston (chapter 1), on the other hand, uses the two terms interchangeably but generally interprets dieback as a symptom leading to decline. Manion (294, 295) attempted to make a clear distinction between dieback and decline. Dieback is a general response to environmental stress. Trees dieback to recover a balance between the water requirements of the crown and the ability of the roots and stem transport system to supply the needs. Dieback in this context is a stabilizing and recovery process. Dieback can also lead to further dieback and decline if the balance is not restored.

The terms dieback and decline are central to much of the confusion on forest decline as an environmental crisis. If forest decline surveys (see examples in this volume) use crown density as the key symptom of decline, they are pooling together the trees that are recovering with those that are deteriorating. Included also are "normal" trees with thin crowns associated with many natural factors and diseased trees with thin crowns because of root, stem or foliage invasion or infestation. Pooled together these appear as an environmental crisis. To appropriately evaluate decline, as contrasted to all of these other problems will require more sophistication than has been currently applied to dieback evaluation.

To avoid the problem Auclair et al. (chapter 3) and others do not use the term decline. Unfortunately this does not solve the problem of pooling normal, deteriorating, and recovering trees. Unless there is some major synchronizing stressing event, the problems associated with more localized and less dramatic events can only be separated based on multiple surveys as described by McLaughlin et al. (chapter 8) and careful diagnostic procedures (chapter 5).

The terms trees and forests should be addressed before going further. Some have difficulty in recognizing that forests are primarily populations of trees and therefore have difficulty in relating a tree decline concept to a forest decline concept. Skelly (chapter 5) interprets forest decline as essentially a misnomer and suggests "forest species decline" as an alternative. This may be interesting semantics, but the real issue is why individuals or segments of the population dieback and decline while others may respond with some twig dieback but eventually recover.

Various authors contrast markedly on which key factors are emphasized. Houston (chapter 1) emphasizes the central importance of organisms of secondary action in the demise of stressed trees, while Muller-Dombois (chapter 2) emphasizes the importance of genetics and aging leading to senescence and vulnerability to stressing agents. A third level of emphasis is presented by Auclair et al. (chapter 3). They highlight low temperature injuries associated with climatic events as the triggering factor. The emphasis on the trigger event provides a larger scale interpretation of why these problems occur at various point in time, but it does not address as clearly the spatial, individual, or populations of trees issues of the Houston and Muller-Dombois concepts. The Manion (295) definition of decline as "an interaction of interchangeable, specifically ordered abiotic and biotic factors to produce a gradual general deterioration, often ending in death of trees" avoids emphasis on key factors by clearly emphasizing the interaction and ordering of the associated factors.

Each of the decline models involve a number of interacting factors but differ on some key points. Houston (chapter 1) expresses the model as a series of stress and response reactions. Manion (291, 295), using the predisposing, inciting, and contributing terms of Sinclair (428), developed a three tiered inward decline spiral model (Fig. 1). Within each ring are a number of stressing factors to indicate the interchangeability of the various factors. Each decline situation involves one factor from each ring. The model is therefore a general diagram for developing a series of different decline scenarios. The Muller-Dombois (chapter 2) enumerated model likewise uses the three steps of the Sinclair (428) model but slightly different terms. Simplified forest structure "s",

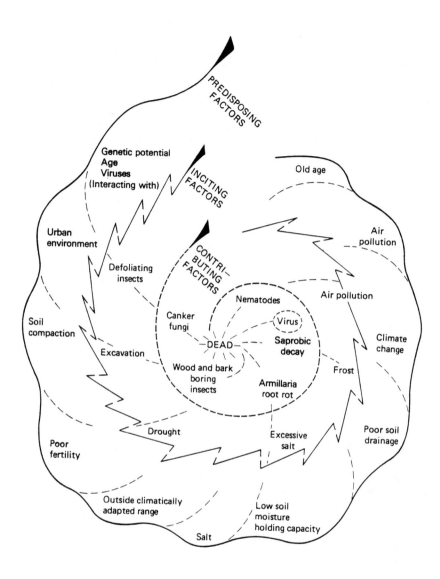

Fig. 1. Decline disease spiral (Reprinted with permission from Tree Disease Concepts by Paul D. Manion c 1991, Prentice-Hall, Inc., Englewood Cliffs, NJ).

edaphically extreme sites "e", periodically recurring perturbations "p" and biotic agents "b" factors are the features of his model. Wallace (486) described an interconnected radiating lines and superimposed circles model in an earlier review article. The distances from the center relate to the level of direct involvement by stressing factors and the interconnections among the rings indicate interactions among the stressing factors.

The similarities among the models sometimes leads to over simplification of the concepts. It is important at this early stage of conceptual development to identify the key points of similarity and dissimilarity. Note for example the contrasting interpretation of the term predisposition and the number of stressing steps among the authors. Consider for example the single agent caused declines of Sinclair and Hudler (430). Note also the role of aging in the various models. These and other features of the decline concept will have to be resolved over time.

SURVEYS OF FOREST DECLINE AND MANAGEMENT RECOMMENDATIONS

A number of chapters provide details and analyses of forest decline surveys. We can provide a few general comments for perspective, but the reader should recognize that we may be too remote from the problem to properly interpret the information.

The German and other European forest health surveys are described and interpreted by Kandler (chapter 4). Crown transparency, the key character, and foliage discoloration are clearly not new symptoms nor are they unique indicators of specific problems. In spite of the limitations, the annual surveys clearly demonstrate no synchrony of changing health conditions of various tree species.

Hennon et al. (chapter 6) detail extensive field evaluation of the Alaska yellow-cedar problem. Note the attempt to identify a single causal factor and the eventual characterization of poorly drained soils as at least one key factor. The collection of symptoms and age distribution of affected trees might suggest that they are potentially mixing more than one problem which therefore limits their ability to finally resolve the issue.

The subject of sugar maple declines, a topic of long

standing interest to forest pathologists of northeastern United States and eastern Canada is addressed in a number of chapters. As suggested by Allen et al. (chapter 7), we all have difficulty with land managers and the general public when we "take exception" to the simplistic explanations of widespread death and destruction of sugar maple associated with acid rain. The magnitude of the maple problem may be better expressed by McLaughlin et al. (chapter 8) when they conclude that surveys of Ontario demonstrate that most of the forests are in the healthy to trace decline classes.

Daoust et al. (chapter 9), using aerial field survey data for Quebec, demonstrate a graphic method of expressing and overlaying the data against other graphic factors. They found little relationship of maple dieback and site quality. The problems of relating single observation crown based survey data that pools declining, recovering, diseased, and healthy trees cannot be resolved through any data modification procedure. As noted earlier, this type of pooling based on visual symptoms may lead to serious confusion. Recurrent observations and detailed diagnoses are required.

Allen et al. (chapter 7) draw on four historical examples and one case study to suggest a key involvement of past harvesting practices in maple declines. These observations lead them to suggest forest management recommendations to reduce declines. Ammon et al. (chapter 10) also suggest that specific silvicultural treatments may reduce decline in southern oaks. One should recognize that these and many other points are judgement rather than analytically derived conclusions.

The survey papers of this volume demonstrate some of the approaches to collecting and manipulating forest decline data. The specific characterization of decline is difficult to standardize. McLaughlin et al. (chapter 8) developed a semi-subjective decline index based on twig dieback, foliar color, and leaf size. Apple and Manion (26) presented a step-wise regression based decline index for Norway maples that identified crown density, crown shape, and small dead limbs as the key variables.

Ammon et al. (chapter 10) avoids the problem of quantifying decline in the field by characterizing decline sites as those on which one oak exhibited terminal defoliation and death. The initial data analysis from such a survey identifies only very dominant factors that may or may not have any functional relationship to the problem of concern. A recent study by Innes and Boswell (223) on the relationships among data from inventories on forest condition concludes by saying that "Although crown density and discoloration provide much useful information, collection of data related to other symptoms is required".

DECLINES AS A NATURAL STABILIZING PROCESS

We have tried to highlight some of the features of Forest Decline Concepts as they have been presented here, but we need to bring in at least one other point that has not been addressed. Are declines a part of the natural background probability of survival models for individual trees? If one calculates annual mortality from the figures on USDA Forest Service inventory plots provided by Teck and Hilt (458), annual losses of 0.5 to 2.8% are demonstrated for trees (12.5 cm or larger) of northeastern forests. These losses, interpreted as self thinning, are contrasted to losses associated with catastrophic features such as fires, windstorms, and epidemics of external agents. Is this really self thinning and therefore not decline (Houston, chapter 1; 295) or is this mortality of mature canopy dominants? The photograph on the cover of the Teck and Hilt publication and most views of mature forests show distinct mortality of canopy dominants.

This mortality (decline induced) of the mature dominant trees was interpreted by Manion (295) as a process of natural selection against the most aggressive competitive genotypes (or species) to balance the losses to competition of the less aggressive genotypes. During periods of uniform weather conditions, competition is the primary mortality factor in forest development. Competitive dominants balance rapid turnover of roots and top to capture resources and build them into a framework that allows exclusive occupation of a site over an extended period of time (156).

In contrast, during unusual climatic events or biologically induced stress periods, the competitive dominant may be most vulnerable. Their large size has stretched the limits of coordination of the uptake, transport, storage, and photosynthetic systems. The ecological penalties associated with emphasis on rapid turnover, based on very active foraging for resources of the competitive dominants, put them at a disadvantage to stress tolerant dominants that utilize a strategy of conservation of limited resources to survive (156). Dramatic injury induces seed production and other responses of mature competitive dominant trees that divert needed recovery resources and further limits the reestablishment of a balanced system. Gradual deterioration and death are hastened by organisms of secondary action. Less competitive subcanopy trees (stress dominants (156)) or a new seedling-based population are allowed to flourish for a while and contribute to the gene pool. The resultant mix over time and space of different types of dominants and genotypes is the foundation of succession, but it also provides a broad genetic base for long-term stability of the forest system.

The current emphasis in community dynamics as a process rather than climax oriented end-point interpretation of an earlier period (355) provide a context for the natural selection decline process. Plant community dynamics is founded on an evolutionary based zero force relationship that specifies the conditions and forces to produce no change in the system (355). Short-term changes occur over time in response to site availability, species availability and species tolerance for the conditions (355). Although most of the current emphasis is on species changes, the concepts can also be applied to genetic changes within species. Likewise, current emphasis on short-term changes need to be interpreted in the context of zero force or no net change in the long-term. Forest decline (death of competitive dominants) may occur in response to short-term events, but the process may be a natural response to maintain long-term equilibrium within and among species.

The assumption that a natural selection process is the driving factor for declines may or may not be correct, but the need to understand the underlying linkage of mortality in this mature canopy dominant population to plant community structure may be critical to understanding forest declines.

SUMMARY

The rational as well as emotional issues of acid rain have demonstrated an essential need to better define and understand forest declines and other forest problems. The current concern with the greenhouse effect and global warming will further focus the need to better define and understand our forest ecosystem. We have to understand why, in the absence of specific pathogens, some (mature?) trees gradually lose vitality and die and others do not, even though general growing conditions appear similar for both. Knowing this, we may be able to theorize and, even explain, why some tree species or forest stands decline while other(s) do not. Predictive models based on a fundamental decline concept (if there is one) could possibly allow us to alter some stand factors to keep or change our forests according to our needs.

As indicated in the Foreword, this volume does not provide a consensus definition of forest decline nor does it express a consensus concept on the role of pollutants in forest decline. In this overview chapter, we have pointed out here only a few of the differences of opinion and key points. Other contributors might have emphasized different aspects. The reader is encouraged to analyze the papers carefully to identify their key points and evaluate the validity and strength of their arguments. An extensive reference list provides a wealth of information and adds value to this volume. These chapters and references represent a foundation for the development of a functional forest decline concept.

LITERATURE CITED

1. Ahrens, Von D., Hanss, A., and Oblander, V. 1988. Bericht über die räumliche verteilung von luftschadstoffen in Sudwestdeutschland. Forstw. Cbl. 107:326-341.
2. Alexander, S.A., and Carlson, J.A. 1989. Visual Damage Survey Project Manual, National Vegetation Survey, USDA-For. Serv. Southeast For. Exp. Sta., Research Triangle Park, NC.
3. Allen, D.C. 1987. Insects, declines, and general health of northern hardwoods: issues relevant to good forest management. Pages 252-285. in: R.D. Nyland (ed.). Managing northern hardwoods - Proceedings of a Silvicultural Symposium. June 1986. SUNY Coll. Environ. Sci. & For., Syracuse. Fac. For. Misc. Pub. No. 13 (SAF Pub. 87-03).
4. Anonymous. 1953. Report of symposium on birch dieback. Summary of the meetings held in Ottawa, Canada, on March 21 & 22, 1952. Part 1 and Part 2. Canada Department of Agriculture, Forest Biology Division, Science Service, Ottawa, Canada.
5. Anonymous. 1964. The causes of maple blight in the Lake States. USDA For. Serv. Res. Paper LS-10. 15 pp.
6. Anonymous. 1964. Studies of maple blight. Res. Bull. 250. University of Wisconsin, Madison, WI. 129 pp.
7. Anonymous. 1980. Mortality atlas of Canada. Health and Welfare Canada, Canada Government Publication Centre, Supply and Services, Hull, Quebec.
8. Anonymous. 1984. The acid rain story. Env. Can., Ottawa, Ont. Canada.
9. Anonymous. 1989. The dying off of the forest. Ministerium fur Ernahrung, Landwirtschaft, Umwelt and Forsten, Baden-Wurttemberg, Freiburg im. Br., FRG.
10. Anonymous. 1985. Cooperative Survey of Red Spruce

and Balsam Fir Decline and Mortality in New York, Vermont and New Hampshire 1984. Forest Service, Northeastern Area, NA-TP-11.
11. Anonymous. 1986. Zweiter Bericht. Forschungsbeirat Waldschäden/Luftverunreinigungen Kernforschungszentrum Karlsruhe GmbH. PF 3640, 7500 Karlsruhe 1.
12. Anonymous. 1987. Daily climatological data: Lennoxville, Quebec, October 1970-May 1987. Canada Department of Environment, Atmospheric Environment Service, Downsview, Ontario, Canada.
13. Anonymous. 1897. Pennsylvania Division of Forestry, Third Annual Report - Part 2. PA Bur. For. Harrisburg, PA.
14. Anonymous. 1987 to 1990. Sanasilva Waldschadensbericht Eidgenössische Anstalt für das forstliche Versuchswesen, CH-8903 Birmensdorf, Switzerland.
15. Anonymous. 1988. Quebec historical streamflow summary to 1986. Canada Department of Environment. Inland Waters Directorate, Water Resources Branch, Water Survey of Canada, DSS Cat. No. EN 36-418/1986-8 Ottawa. Canada.
16. Anonymous. 1988. International cooperation on assessment and monitoring of air pollution effects on forests. Economic Commission for Europe. Minutes of the fourth meeting of the program task force held on 3-4 May 1988 in Koenhigswinter, FRG.
17. Anonymous. 1988. Forest damage surveys in Europe. United Nations Economic and Social Council, Executive summary document EB.AIR/R39, Rev. Economic Commission for Europe.
18. Anonymous. 1988. SAS/STAT User's Guide, Release 6.03 Edition. SAS Institute, Inc., Cary, NC.
19. Anonymous. 1989. Luft, Chemie und Umwelt. December 1989. Verband der Chemischen Industrie e.V. Frankfurt a.M., FRG.
20. Anonymous. 1989. TECPLOT: An interactive plotting program. Amtec Engineering, Bellevue, WA.
21. Anonymous. 1989. Dritter Bericht. Forschungsbeirat Waldschäden/Luftverunreinigungen Kernforschungszentrum Karlsruhe GmbH. PF 3640, 7500 Karlsruhe 1.

22. Anonymous. 1989. Information zum waldsterbelehrpfad Schauinsland, Arbeitskreis Wald, Bund fur Umwelt and Naturschutz, Freiburg, FRG (Mimeo).
23. Anonymous. 1990. Air Pollution and Forest Decline: Is There a Link? USDA-For. Serv. Ag. Inf. Bull. 595. Washington, DC. 13 pp.
24. Anonymous. 1990. Acidic deposition: State of science and technology. Rept. on the NAPAP Int'l. Conf., Feb. 11-16. Hilton Head, SC. NAPAP Office of the Director, Washington, DC. 98 pp.
25. Appel, D.N. and Stipes, R.J. 1984. Canker expansion on water-stressed pin oaks colonized by Endothia gyrosa. Plant Dis. 68:851-853.
26. Apple, J. D. and Manion, P. D. 1986. Increment core analysis of declining Norway maples. Urban Ecology 9:309-321.
27. Arnold, R.H. 1967. A canker and foliage disease of yellow birch. I. Description of the causal fungus, Diaporthe alleghaniensis sp. nov., and the symptoms on the host. Can. J. Bot. 45:783-801.
28. Auclair, A.N.D. 1989. Climate change theory of forest decline. Pages 1-30 in: Proceedings IUFRO Conference on Woody Plant Growth in a Changing Physical and Chemical Environment. D.P. Lavender, ed. July 27-31, 1987, University of British Columbia, Vancouver, British Columbia, Canada.
29. Auclair, A.N.D. 1989. Winter thaw-freeze as a mechanism of forest dieback. Pages 171-172 in: Poster Abstracts Vol. 1, International Congress on Forest Decline Research, State of Knowledge and Perspectives, October 2-6, 1989, Friedrichshafen, West Germany.
30. Auclair, A.N.D. 1989. An historical overview of forest decline in eastern Canada. In: Proceedings of the 70th Annual Meeting of the Woodlands Section, Canadian Pulp and Paper Association March 21-22, 1989, Montreal, Quebec, Canada.
31. Auclair, A.N.D. 1989. Cavitation as a mechanism of dieback in Northern Hardwoods. in: Proc. Récherche sur le dépérissement des erablieres, Centre de récherche acéricole, ministère de l'Agriculture, des Pêcheries et de l'Alimentation, Quebec. 5 pp.
32. Auclair, A.N., Martin, J.C., and Walker, S.L. 1990. A case study of forest decline in western Canada and the

adjacent United States. J. Water Air Soil Pollut., 53:13-31.
33. Bach, W. 1985. Waldsterben: Our dying forests - Part III. Forest Dieback: Extent of damages and control strategies. Experimentia 41:1095-1104.
34. Baker, W.L. 1941. Effect of gypsy moth defoliation on certain forest trees. J. For. 39:1017-1022.
35. Balakrishnan, N. and Mueller-Dombois, D. 1983. Nutrient studies in relation to habitat types and canopy dieback in the montane rain forest ecosystem, Island of Hawaii. Pac. Sci. 37:339-350.
36. Balch, R.E. 1927. Dying oaks in the southern Appalachians. For. Worker 7:13.
37. Balch, R.E. 1944. The dieback of birch in the Maritime region. Contrib. No. 3. Ottawa: Can. Dep. Agric., Sci. Serv., Entom. Div.; Mimeo.
38. Ballach, Von H.J., and Brandt, C.J. 1983. BML-Forest decline inventory 1983 analysis of causes without differential diagnosis? Staub-Reinhalt. 43:448-452.
39. Barnard, J.E. 1989. Environmental health concerns: a role for forest inventory and monitoring, in: PRoc. IUFRO Conf. State of the Art Methodology of Forest Inventory. Syracuse, NY (in press).
40. Barnard, J.E., Lucier, A.A., Johnson, A.H., Brooks, R.T., Karnowsky, D.F., and Dunn, P.H. 1990. Acid deposition state of science and technology, changes in forest health and productivity in the United States and Canada. Natl. Acid Precip. Assessment Prog., SOS/T Rep. No. 16. 186 pp.
41. Barter, G.W. 1953. Work on the birch dieback problem in the Maritime region. Pages 12-21 in: Report of the Symposium on Birch Dieback. can. Dep. of Agric. Ottawa, Canada.
42. Barter, G.W. 1953. Some characteristics of dieback. Pages 22-23 in: Report of the Symposium on Birch Dieback. Can. Dep. Agric., Ottawa, Canada.
43. Barter, G.W. 1957. Studies of the bronze birch borer, Agrilus anxius Gory, in New Brunswick. Can. Entomol. 89:12-36.
44. Barter, G.W. and R.E. Balch. 1950. On the apparent spread of birch "dieback". Dom. Dep. Agric., Bi-monthly Progress Rep. 6(3).
45. Bassett, E.N. and Fenn, P. 1984. Latent colonization

and pathogenicity of Hypoxylon atropunctatum on oak. Plant Dis. 68:317-719.
46. Bassett, E.N., Fenn, P., and Mead, M.A. 1982. Drought-related oak mortality and incidence of hypoxylon canker. Arkansas Farm Res. 31:(1)8.
47. Bauce, E., and Allen, D.C. 1991. Etiology of a sugar maple decline. Can. J. For. Res. 21:686-693.
48. Bauer, F. 1985. Die Sache mit dem Wald. BLV Verlagsgesellschaft, München-Wien-Zürich.
49. Bazire, P. 1987. *Results of the Survey of Health Conditions of the French Forest in 1986.* in: Forest Decline and Reproduction: Regional and Global Consequences. L. Kairiukstis, S. Nilsson and A. Straszak eds. International Institute for Applied Systems Analysis, Laxenburg, Austria, 387-397.
50. Beal, J.A. 1926. Frost kills oak. J. For. 24:949-950.
51. Becker, M. 1987. Bilan de santé actuel: et retrospectif du sapin (Abies alba Mill.) dans les Vosges. Étude écologique et dendrochronologique. Ann. Sci. For. (Paris) 44:379-409.
52. Becker, M. 1989. The role of climate on present and past variability of silver fir forest in the Vosges mountains of northeastern France. Can. J. For. Res. 19:1110-1117.
53. Becker, M., Bräker, O.-U., Kenk, G., Schneider, O., and Schweingruber, F.-H. 1990. Kronenzustand und Wachstum von Waldbäumen im Dreiländereck Deutschland-Frankreich-Schweiz in den letzten Jahrze hnten. Allgem. Forst. Z. 45:263-274.
54. Becker, M., Bräker, O.U., Kenk, G., Schneider, O., and Schweingruber, F.H. 1990. Appearance of crowns and growthof trees in the last few decades in the border regions of Germany, France and Switzerland. Rev. For. Fr. 42:284-300.
55. Becker, M., Landmann, G., and Levy, G. 1989. Silver fir decline in the Vosges Mountains (France): Role of climate and silviculture. J. Water Air Soil. Pollut. 48:77-86.
56. Berbee, J.G. 1957. Virus symptoms associated with birch dieback. Can. Dep. Agric., Sci. Serv., Forest Biol. Division, Bi-Monthly Progress Rept. 13(1)51.
57. Bernier, B. and Brazeau, M. 1988. Foliar nutrient status in relation to sugar maple decline in the Quebec

Appalachians. Can. J. For. Res. 18:754-761.
58. Berry, F.H. 1955. Investigation of possible causes of sweet gum blight. Plant Dis. Rep. 39:270-272.
59. Bier, J.E. 1959a. The relation of bark moisture to the development of canker diseases caused by facultative parasites. I. <u>Cryptodiaporthe</u> canker of willow. Can. J. Bot. 37:229-238.
60. Bier, J.E. 1959b. The relation of bark moisture to the development of canker diseases caused by native, facultative parasites. II. <u>Fusarium</u> canker on black cottonwood. Can. J. Bot. 37:781-788.
61. Bier, J.E. 1961a. The relation of bark moisture to the development of canker diseases caused by native, facultative parasites. IV. Pathogenicity studies of <u>Cryptodiaporthe salicella</u> (Fr.) Petrak and <u>Fusarium lateritium</u> Nees, on <u>Populus trichocarpa</u> Torren and Gray, <u>P.</u> "robusta", <u>P. tremuloides</u> Michx., and <u>Salix</u> sp. Can. J. Bot. 39:139-144.
62. Bier, J.E. 1961b. The relation of bark moisture to the development of canker disease caused by native, facultative parasites. VI. Pathogenicity studies of <u>Hypoxylon pruinatum</u> (Klotsch) Cke., and <u>Septoria musiva</u> Pk. on species of <u>Acer</u>, <u>Populus</u> and <u>Salix</u>. Can. J. Bot. 39:1555-1561.
63. Bier, J.E. 1964. The relation of some bark factors to canker susceptibility. Phytopathology 54:250-253.
64. Binns, W.O. and Redfern, D.B. 1983. Acid rain and forest decline in West Germany. Forestry Commission Res. and Dev. Paper 131. For. Com. Res. Sta., Alice Holt Lodge, Farnham, Surrey, England. 13 pp.
65. Binns, W.O., Redfern, D.B., and Reynolds, K. 1987. Forest Decline - the View from Britain. in: NATO ASI Series, Vol. G16. Effects of Atmospheric Pollutants on Forests, Wetlands and Agricultural Ecosystems. T.C. Hutchinson and K.M. Meema ed. Springer-Verlag, Berlin.
66. Bird, T., Kile, G.A., and Podger, F.D. 1975. The eucalypt crown diebacks - A growing problem for forest managers. Aust. For. 38:173-187.
67. Blank, L.W. 1985. A new type of forest decline in Germany. Nature (London) 314:311-314.
68. Blank, L.W., Robert, T.M., and Skeffington, R.A. 1988. New perspectives on forest decline. Nature 336:27-30.

69. Boas, F. 1942. Dynamische Botanik. Lehmanns Verlag München.
70. Bordeleau, C. 1987. La carte des relevés aériens du dépérissement des érablières au Québec en 1985 et 1986. Mimeo.
71. Bosch, C., Pfannkuch, E., Baum, U., and Rehfuess, K.E. 1983. über die Erkrankung der Fichte (Picea abies KARST.) in den Hochlagen des Bayerischen Waldes. Forstw. Cbl. 102:167-181.
72. Braathe, P. 1957. Is there a connection between the birch dieback and the March thaw of 1936? For. Chron. 33:358-363.
73. Brechtel, H.M., Dieterle, G., Innes, J.L., Krause, G.H.M., Materna, J., Thomson, M.G., and Scholz, F. 1990. Interim report on cause-effect relationships in forest decline. Global Environment Monitoring Systems. Convention on Long-Range Transboundary Air Pollution. For. Commission. Alice Holt Lodge. Farnham-Surrey England. 173 pp.
74. Brokaw, N.V.L. 1985. Treefalls, regrowth, and community structure. Pages 53-69 in: The Ecology of Natural Disturbance and Patch Dynamics. S.T.A. Pickett and P.S. White, eds. Academic Press, Inc., New York.
75. Bruck, R.I. 1989. Survey of diseases and insects of Fraser fir and red spruce in the southern Appalachian Mountains. Eur. J. For. Path. 19:389-398.
76. Bruck, R.I., and W.P. Robarge. 1988. Change in forest structure in the boreal ecosystem of Mount Mitchell, North Carolina. Eur. J. For. Path. 18:357-366.
77. Bruck, R.I., W.P. Robarge, and A. McDaniel. 1989. Forest decline in the boreal montane ecosystems of the southern Appalachian mountains. J. Water, Air, Soil Pollut. 48:161-180.
78. Bucher, J.B. 1987. Forest Damage in Switzerland, Austria and Adjacent Parts of France and Italy in 1984. Page 43-58 in: NATO ASI Series, Vol. G16. Effects of Atmospheric Pollutants on Forests, Wetlands and Agricultural Ecosystems. T.C. Hutchinson and K.M. Meema (eds.), Springer-Verlag, Berlin.
79. Bucher, J.B. 1989. Zur Diagnose der neuartigen Waldschäden ("Waldsterben") in der Schweiz. in: Sanasilva Tagungsbericht: Kritische Analyse des

Kenntnisstandes in Sachen Ursachenforschung Waldschäden. 5 April 1989, ETH-Zentrum Zürich.
80. Bucher, J.B., and Bucher-Wallin, I. eds. 1989. Air pollution and forest decline. Proc. 14th Int'l. Meeting for Specialists in Air Pollution Effects on Forest Ecosystems, IUFRO, P 2.05, Interlaken, Switzerland, Birmensdorf.
81. Burger, H. 1953. Holz, Blattmenge und Zuwachs. XIII. Mitteilung. Fichten im gleichaltrigen Hochwald. Mitt. EAFV 29, 38-130.
82. Burgess, A.F. 1922. Discussion. In the gypsy moth, an imminent menace to the forst and shade trees of the state of New York. NY State Department of Farms and Markets Agricultural Bulletin 148:33-41.
83. Burton, P.J., and D. Mueller-Dombois. 1984. Response of <u>Metrosideros polymorpha</u> seedlings to experimental canopy opening. Ecology 65:779-791.
84. Campbell, W.A., and O.L. Copeland. 1954. Littleleaf disease of shortleaf and loblolly pines. U.S. Dep. Agric. Circ. 940. 41 pp.
85. Cannell, M.G.R., and L.J. Sheppard. 1982. Seasonal changes in the forst hardiness of provenances of <u>Picea sitchensis</u> in Scotland. Forestry 55:137-153.
86. Cape, J.N., I.S. Patterson, and J. Wolfenden. 1989. Regional variation in surface properties of Norway spruce and scots pine needles in relation to forest decline. Environ. Pollut. 58:325-342.
87. Carey, A.C., Miller, E.A., Geballe, G.T., Wargo, P.M., Smith, W.H., and Siccama, T.G. 1984. <u>Armillaria mellea</u> and decline of red spruce. Plant Dis. 68:794-795.
88. Carrier, L. 1990. Personal communication. Direction de la recherche et du dévelopement, Ministère de l'Energie et des Ressources, Québec. in: S.L. Walker, and A.N.D. Auclair eds. Forest declines in eastern Canada. Federal LRTAP Liaison Office Report, Atmospheric Environment Service, Canada Dept. Environment, Downsview, Ontario, Canada.
89. Castello, J.D. 1989. Viruses in forested ecosystems: Inciting and predisposing factors in tree decline. pp. 351-357 in: Proceedings International Congress on Forest Decline Research: State of knowledge and perspectives. B. Ulrich, ed. Kernforschungszentrum

Karlsruhe GmbH. PF 3640, 7500 Karlsruhe 1, FRG.
90. Cehak, K. 1985. Niederschlagschemische Beobachtungen in wenig industrialisierten Gebieten Österreichs. Pages 295-312 in: VDI-Kommission Reinhaltung der Luft, ed., Waldschäden, Einflußfaktoren und ihre Bewertung. VDI-Berichte 560. VDI-Verlag GmbH Düsseldorf, FRG.
91. Chamberlin, T.C. 1987. Studies for students: The method of multiple working hypothesis. J. Geology V:837-848.
92. Chapman, R.N. 1915. Observations on the life history of Agrilus bilineatus. J. Agr. Res. 3:283-293.
93. Clark, J., and Barter, G.W. 1958. Growth and climate in relation to dieback of yellow birch. For. Sci. 4:343-364.
94. Claus, J. 1928. Das Tannensterben im deutschen Walde. Sudetendeutsche Forst-Jgdz. 28:279-281.
95. Clements, F.E. 1928. Plant succession and indicators. Republ. 1963. Hafner Publ. Co., New York. 453 p.
96. Cook, E.R., Johnson, A.H., and Blasing, T.J. 1987. Forest decline: modeling the effect of climate in tree rings. Tree Physiol. 3:27-40.
97. Cotter, H.V.T., and Blanchard, R.O. 1981. Identification of the two Nectria taxa causing bole cankers on American beech. Plant Dis. 65:332-334.
98. Cowling, E.B. 1985. Comparison of Regional Declines of Forests in Europe and North America: A Possible Role of Airborne Chemicals. Pages 217-234 in: Air Pollutants Effects on Forest Ecosystems. May 8-9, St. Paul MN. The Acid Rain Foundation.
99. Cowling, E.B. 1989. Individual comments and suggestions deriving from the lectures, discussion sessions, posters and informal discussions at the International Congress in Friedrichshafen. Pages 939-943 in: Proceedings International Congress on Forest Decline Research: State of knowledge and perspectives. Ulrich, B., ed. Kernforschungszentrum Karlsruhe GmbH. Postfach 3640, 7500 Karlsruhe 1, FRG.
100. Cramer, H.H. 1984. Über die Disposition mitteleuropäischer Forsten für Waldschäden. Pflanzenschutz-Nachr. Bayer 37:97-207.
101. Cramer, H.H. 1985. Waldschäden in Geschichte und Kunstgeschichte. Bonner Universitätsblätter 1985:37-

45.
102. Cramer, H.H., and Cramer-Middendorf, M. 1984. Untersuchungen über Zusammenhänge zwischen Schapensperioden und Klimafakotren in mitteleuropäischen Forsten seit 1851. Pflanzenschutz-Nachr. Bayer 37:208-234.
103. Cressie, N. 1991. Statistics for spatial data. John Wiley and Sons, NY.
104. Crist, C.R., and Schoeneweiss, D.F. 1975. The influence of controlled stresses on susceptibility of European white birch stems to attack by Botryosphaeria dothidea. Phytopathology 65:369-373.
105. Daoust, G., Ansseau, C., Granberg, H.B., and Van Hulst, R. 1990. Réflectance spectrale de feuilles d'érable à sucre (Acer saccharum) souffrant de dépérissement. Proceedings Colloque sur le dépérissement des érablières: Causes et solutions possibles. May 14-17, 1990, Université Laval, Québec.
106. Davaiault, L. 1953. Evidence of spread or intensification as indicated by location and time of injury. Pages 42-49 in: Report of the Symposium on Birch Dieback. Canada Department of Agriculture, Ottawa, Ontario, Canada.
107. Davidson, E.M. 1988. The role of waterlogging and Phytophthora cinnamoni in the decline and death of Eucalyptus marginata in western Australia. GeoJournal 17:239-244.
108. Davidson, E.M., and Tay, F.C.S. 1985. The effect of waterlogging on seedlings of Eucalyptus marginata. New Phytol. 101:743-753.
109. DeGroot, R.C. 1965. A dieback of Fagus grandifolia associated with Lepidosaphes ulmi infestations in New York. Phytopathology 55:1055.
110. DeHayes, D.H., Ingle, M.A., and Waite, C.E. 1989. Developmental cold tolerance of red spruce and potential perturbations from natural and anthropogenic factors. Pages II-1 to II-10 in: Proceedings Workshop on air pollution and winter injury of red spruce; Edinburgh, Scotland; 1989 April 17-18.
111. Delatour, C. 1983. Les dépérissements des chenes on Europe. Rev. For. Franc. 35:265-268.
112. Devevre, O. 1990. Mise en evidence experimentale d'une microfole rhizospherique deletere associee au

deperissement de l'epicea en France et en Allemagne. Thesis Universite de Nancy I, France.
113. Doane, C.C., and McManus, M.L., eds. 1981. The gypsy moth: Research toward integrated pest management. U.S. Dep. Agric. Tech. Bull. 1584. Washington, DC. 529 p.
114. Donaubauer, E. 1987. Auftreten von Krankheiten und Schädlingen der Eiche und ihr Bezug zum Eichensterben. österr. Forstzeitung 98:46-48.
115. Dull, C.W. 1982. Assessment of timber mortality on the Lee Ranger District, George Washington National Forest, Virginia. Rep. 83-3-11. Athens, GA: U.S. Department of Agriculture Forest Service, State and Private Forestry, Southeastern Area. 17 pp.
116. Dull, C.W., Ward, J.D., Brown, H.D., Ryan, G.W., Clerke, W.H., and Uhler, R.J. 1988. Evaluation of spruce and fir mortality in the southern Appalachian Mountains. USDA-For. Serv. Prot. Rept. R-8-PR-13.
117. Dunbar, D.M., and Stephens, G.R. 1975. Association of twolined chestnut borer and shoestring fungus with mortality of defoliated oak in Connecticut. For. Sci. 21:169-174.
118. Ehrlich, J. 1934. The beech bark disease. A Nectria disease of Fagus, following Cryptococcus fagi (Baer.). Can. J. Res. 10:593-692.
119. Englund, E., and Sparks, A. 1988. GEO-EAS (Geostatistical environmental assessment software) User's Guide. Environmental Monitoring Systems Laboratory, Office of Research and Development, U.S.E.P.A., Las Vegas, NV.
120. Estivalet, D., Perrin, R., Le Tacon, F., and Bouchard, D. 1990. Nutrition and microbiological aspects of decline in the Vosges forest area (France). For. Ecol. and Mgmt. 37:233-248.
121. Eyre, F.H. ed. 1980. Forest cover types of the United States and Canada. Soc. Am. For., Wash., DC. 148 pp.
122. Farrell, M.P. 1987. Master index for the carbon dioxide research state-of-the-art report series. Report/DOE ER-0316, Oak Ridge National Laboratory, Oak Ridge, TN.
123. Fedde, G.F. 1964. Elm spanworm (Ennomos subsignarius), a pest of hardwood forests in the southern Appalachians. J. For. 62:102-106.

124. Feger, K.H. 1988. Historical changes in catchment use. 1988. COST Workshop in Effects of Land Use in Catchments on the Acidity and Ecology of Natural Surface Waters. CEC, Univ. of Wales Inst. of Sci. and Tech. Cardiff. 10 pp.
125. Fergus, C.L., and Ibberson, J.E. 1956. An unexplained extensive dying of oak in Pennsylvania. Plant Dis. Reporter 40:748-749.
126. Fink, S., and Braun, H.J. 1978. Zur epidemischen Erkrankung der Weiβtanne Abies alba Mill, I. Untersuchungen zur Symptomatik und Formulierung einer Virus-Hypothese. Allg. Forst Jagdztg. 149:145-150.
127. Fosberg, F.R. 1948. Derivation of the flora of the Hawaiian Islands. Pages 107-199 in: Insects of Hawaii, vol. 1, E.C. Zimmerman, ed. Univ. Hawaii Press, Honolulu.
128. Foukal, P.V. 1990. The variable sun. Sci. Am. 262(2):34-41.
129. Fowells, H.A. 1965. Silvics of forest trees of the United States. agriculture Handbook No. 271. United States Department of Agriculture, Forest Service. Washington, DC.
130. Franklin, J.F. 1988. Importance and justification of long-term studies in ecology. Pages 3-19 in: Long-Term Studies in Ecology, Approaches and Alternatives. G.E. Likens ed. Springer-Verlag, NY.
131. Franklin, G.F., and Dyrness, C.T. 1973. Natural vegetation of Oregon and Washington. USDA For. Serv. Gen. Tech. Rep. PNW-8. Pac. Northwest For. and Range Exp. Sta., Portland, OR 417 pp.
132. Franz, F. 1983. Auswirkungen der Walderkrankungen auf Struktur und Wuchsleistung von Fichtenbeständen. Forstwiss. Centralbl. 102:186-200.
133. Frear, S.M. 1982. What's killing the Alaska yellow-cedar? Amer. For. 88(11):41-43, 62-63.
134. Friedland, A.J., Gregory, R.A., Karenlampi, L., and Johnson, A.H. 1984. Winter damage to foliage as a factor in red spruce decline. Can. J. For. Res. 14:963-965.
135. Gagnon, G. 1987. Situation du dépérissement dans le parcelles d'étude du Ministère d'Energie et Ressources en 1987. Recherche sur le dépérissement des

érablières, Atelier 25-26 November, 1987, Victoriaville. Ministère de l'Agriculture des Pêcheries et de l'Alimentation, Québec, Québec, Canada.

136. Gagnon, G. 1990. Dépérissement des érablières. Forêt Conservation. Suppl. 57:8-10.

137. Gagnon, G., and Bordeleau, C. 1989. Dépérissement des érablières. Pages 8-10 in C. Bordeleau, D. Guerin, L. Innes, and D. Lachance. Quebec 1989 Insectes et Maladies des Arbres; Rapport Annuel. Forêts Canada, Sainte-Foy, Québec, Canada.

138. Gagnon, G. and Roy, G. 1989. Le dépérissement de la forêt feuillue au Québec. Formation COntinue 10:1-10. In: L'Aubelle no. 71. Gamma Design Software. GS+: Geostatistics for the Agronomic and Biological Sciences. Plainwell, MI.

139. Gamma Design Software. GS+: Geostatistics for the Agronomic and Biological Sciences. Plainwell, MI.

140. Games, W. 1963. Mycelium radicis atrovirens in forest soils, isolation from soil microhabitats and identification. Pages 176-182 in: Soil Organisms. J. Doekoen and J. vander Drift, eds. North-Holland Publishing Co., Amsterdam, Holland.

141. Gansner, D.A., Birch, T.W., Arner, S.L., and Zarnoch, S.J. 1990. Cutting disturbance on New England timberlands. North. J. Appl. For. 7:118-120.

142. Gardner, M. 1985. Perpetual motion. Science Digest 93:68-73.

143. Garner, J.H.B., Pagano, T., and Cowling, E.B. 1989. An evaluation of the role of ozone, acid deposition and other airborne pollutants in the forests of eastern North America. USDA-Forest Service, S.E. Exp. Sta. Gen'l. Tech. Rept. SE-59, Res. Tri. Park, NC, 172 pp.

144. Gatsuk, L.E., Smirnova, O.V., Vorontzova, L.B., and Zhukova, L.A. 1980. Age states of plants of various growth forms: a review. J. Ecol. 68:675-696.

145. Gerrish, G. 1988. An explanation of natural dieback based on the "pipe model" analogy. GeoJ. 17:295-299.

146. Giese, R.C., Houston, D.R., Benjamin, D.M., Kuntz, J.E., Kapler, J.E., and Skilling, D.D. 1964. Studies of maple blight. University Wisconsin Madison, Wisconsin Agric. Res. Bull. 250. 128 pp.

147. Gillespie, W.H. 1956. Recent extensive mortality of scarlet oak in West Virginia. Plant Dis. Rep. 40:1121-

1123.
148. Gizzum, W.J., McLaughlin, D.L., and Dixon, M.J. 1990. Results of Ontario hardwood forest health surveys. Poster Session, 19th International Union of Forest Research Organization (IUFRO) World Congress, August 5-11, 1990, Montreal, Quebec, Canada.
149. Glerum, C. 1985. Frost hardiness of coniferous seedlings: principles and applications. in: Evaluating Seedling Quality: Principle, Procedures, and Predictive Abilities of Major Tests. Duryea, M.L., ed. Proceedings of Workshop held October 16-18, 1984. Forest Research Laboratory, Oregon State University, Corvallis, Oregon.
150. Graebner, P. 1924. Degeneration. Pages 48-54 in: Sorauers Handbuch der Pflanzenkrankheiten. Appel, Vol. I, P. Graebner and L. Reh eds. Paul Parey Verlag, Berlin.
151. Graham, S.A. 1965. Causes leading to insect outbreaks. Pages 3-14 in: Insects in Southern Forests. C.B. Martin ed. Louisiana St. Univ. Press, Baton Rouge, LA.
152. Graser, H. 1939. Die Tanne und das Tannensterben. Sächsische Heimatschutzhefte 1939 (5/8):139-148.
153. Greenidge, K.N.H. 1951. Dieback: A disease of yellow birch (Betula lutea Michx.) in eastern Canada. Ph.D. Thesis Department of Biology, Harvard University, Boston, Massachusetts. 300 pp.
154. Griffin, H.D. 1965. Maple dieback in Ontario. For. Chron. 41:295-300.
155. Grimble, D.G., and Kasile, J.D. 1974. A sequential sampling plan for saddled prominent eggs. AFRI Res. Rep. No. 15, St. Univ. Coll. Environ. Sci. and For., Syracuse, NY. 15 pp.
156. Grime, J.D. 1987. Dominant and subordinate components of plant communities: implications for succession, stability and diversity. Page 413-428 in: Colonization Succession and Stability, The 26th British Ecological Society Symposium. A.J. Gray, M.J. Crowley and P.J. Edwards, eds. Blackwell Scientific Publications, Boston.
157. Hamburg, S.P., and Cogbill, C.V. 1988. Historical decline of red spruce populations and climatic warming. Nature 331:428-431.
158. Hamilton, T.D. 1965. Alaskan temperature fluctuations

and trends: an analysis of recorded data. Arctic 18:105-117.

159. Hamm, P.B., Hansen, E.M., Hennon, P.E., and Shaw, C.G. III. 1988. Pythium species from forest and muskeg areas of southeast Alaska. Trans. Br. Mycol. Soc. 91:385-388.

160. Hansbrough, J.R. 1953. The significance of the fungi and viruses associated with birch dieback. Pages 128-35 in: Report of the symposium on birch dieback. Ottawa, Canada: Canadian Department of Agriculture.

161. Hansbrough, J.R., and Stout, D.C. 1947. Viruslike symptoms accompanying birch dieback. Plant Dis. Rep. 31:327.

162. Hansen, E.M., and Hamm, P.B. 1988. Phytophthora species from remote forests of western North America. Phytopathology 78:1519.

163. Hansen, E.M., Hamm, P.B., Shaw, C.G. III, and Hennon, P.E. 1988. Phytophthora drechsleri in remote areas of southeast Alaska. Trans. Br. Mycol. Soc. 91:379-388.

164. Hanson, J.B., Hoffard, W.H., and Orr, P.W. 1976. Forest pest conditions in the Northeast 1975. Upper Darby, PA: USDA, For. Serv., State and Private Forestry, Northeastern area. 25 pp.

165. Hansen, J.E., and Lebedeff, S. 1987. Global trends on measured surface air temperature J. Geophys. Res. 92 (D11):13345-13372.

166. Harcombe, P.A. 1987. Tree life tables. BioScience 37:557-568.

167. Harper, J.L. 1977. Population biology of plants. Acad. Press, NY.

168. Harrington, T.C. 1986. Growth decline of wind-exposed red spruce and balsam fir in the White Mountains. Can. J. For. Res. 16:232-238.

169. Harris, A.S. 1971. Alaska yellow-cedar. USDA For. Serv., Am. Woods-FS 224. 7 pp.

170. Harris, A.S., and Farr, W.A. 1974. The forest ecosystem of southeast Alaska. 7. Forest ecology and timber management. USDA For. Serv. Pacific Northwest Forest and Range Exp. Sta. Portland, OR. Gen. Tech. Rep. PNW-25. 109 pp.

171. Harris, A.S., Hutchison, O.K., Meehan, W.R., Swanston, D.N., Helmers, A.E., Hendee, J.C., and Collins, T.M.

1974. The forest ecosystem of southeast Alaska. 1. The setting. USDA For. Serv. Pacific NW For. and Range Exp. Sta. Portland, OR. Gen. Tech. Rep. PNW-12. 40 pp.

172. Hartley, C., and Merrill, T.C. 1915. Storm and drought injury to foliage of ornamental trees. Phytopathology 5:20-29.

173. Hartmann, G., Nienhaus, F., and Butin, H. 1988. Farbatlas waldschaden: Diagnose von baumkrankheiten, Eugen Ulmer GmbH. & Co., Stuttgart, FRG. 256 pp.

174. Hartmann, G., Blank, R., und Lewark, S. 1989. Eichensterben in Norddeutschland-Verbreitung, Schadbilder, mögliche Ursachen. Forst und Holz 44:475-487.

175. Hawboldt, L.S. 1952. Climate and birch dieback. Bulletin No. 6. Department of Lands and Forests. Province of Nova Scotia, Halifax, Nova Scotia, Canada.

176. Hawboldt, L.S., and Skolko, A.J. 1948. Investigations of yellow birch dieback in Nova Scotia in 1947. J. For. 46:659-671.

177. Heichel, G.H., and Turner, N.C. 1983. CO_2 assimilation of primary and regrowth foliage of red maple (Acer rubrum L.) and red oak (Quercus rubra L.): response to defoliation. Oecologia 57:14-19.

178. Heinselman, M.L. 1981. Fire intensity and frequency as factors in the distribution and structure of northern ecosystems. Pages 7-57 in: Fire Regimes nd Ecosystem Properties. H.A. Mooney, T.M. Bonnickesen, N.L. Christensen, J.E. Lotan and W.R. Reiners, eds. USDA For. Serv. Gen. Tech. Rept. WO. 26. 593 pp.

179. Hennon, P.E. 1986. Pathological and ecological aspects of decline and mortality of Chamaecyparis nootkatensis in southeast Alaska. Ph.D. Thesis. Department of Botany and Plant Pathology, Oregon State University, Corvallis, Oregon. 279 pp.

180. Hennon, P.E. 1990. Fungi on Chamaecyparis nootkatensis. Mycologia 82:59-66.

181. Hennon, P.E., Hansen, E.M., and Shaw, C.G. III. 1990. Causes of basal scars on Chamaecyparis nootkatensis in southeast Alaska. Northwest Sci. 64:45-54.

182. Hennon, P.E., Hansen, E.M., and Shaw, C.G. III. 1990. Dynamics of decline and mortality of Chamaecyparis nootkatensis in southeast Alaska. Can. J. Bot. 68:651-

662.
183. Hennon, P.E., G.B. Newcomb, C.G. Shaw, III, and Hansen, E.M. 1986. Nematodes associated with dying Chamaecyparis nootkatensis in Southeastern Alaska. Plant Dis. 70:352.
184. Hennon, P.E., Shaw, C.G. III, and Hansen, E.M. 1990. Dating decline and mortality of Chamaecyparis nootkatensis in southeast Alaska. For. Sci. 36:502-515.
185. Hennon, P.E., Shaw, C.G. III, and Hansen, E.M. 1990. Symptoms and fungal associations of declining Chamaecyparis nootkatensis in southeast Alaska. Plant Dis. 74:267-272.
186. Hepting, G.H. 1985. A Southwide survey for sweetgum blight. Plant Dis. Rep. 39:261-265.
187. Hepting, G.H. 1963. Climate and forest diseases. Annu. Rev. Phytopathol. 1:31-50.
188. Hepting, G.H. 1971. Diseases of forest and shade trees of the United States. USDA For. Serv. Agric. Handb. 386. 658 pp.
189. Hewitt, C.N., Lucas, P., Wellburn, A.R., and Fall, R. 1990. Chemistry of ozone damage on plants. Chem. and Industry. 15:478-481.
190. Hibben, C.R. 1959. Relations of Stegonosporium ovatum (Pers. ex. Merat) Hughes with dieback of sugar maple (Acer saccharum Marsh.) M.S. Thesis, Cornell Univ. Ithaca, NY. 63 pp.
191. Hibben, C.R. 1962. Investigations of sugar maple decline in New York woodlands. Ph.D. Thesis, Cornell Univ. Ithaca, NY. 301 pp.
192. Hibben, C.R. 1964. Identity and significance of certain organisms associated with sugar maple decline in New York woodlands. Phytopathology 54:1389-1392.
193. Hibben, C.R. 1966. Transmission of a ringspot-like virus from leaves of white ash. Phytopathology 56:323-325.
194. Hibben, C.R., and Bozarth, R.F. 1972. Identification of an ash strain of tobacco ringspot virus. Phytopathology 62:1023-1029.
195. Hibben, C.R., and Silverborg, S.B. 1978. Severity and causes of ash dieback. J. Arbor. 4:274-279.
196. Hibben, C.R., and Wolanski, B. 1971. Dodder transmission of a mycoplasma from ash witches'-broom. Phytopathology 61:151-156.

197. Hinrichsen, D. 1986. World Resources. Basic Books, D. Hinrichsen ed., New York.
198. Hinrichsen, D. 1987. The forest decline enigma. BioScience 37:452-546.
199. Hiruki, C., ed. 1988. Tree mycoplasmas and mycoplasma diseases. The University of Alberta Press. Edmonton, Alberta, Canada.
200. Hiss, W. 1922. Vom Tannensterben. Allg. Forst. Jagdztg. 98:30-33
201. Hitschold. 1934. Das Fichtensterben in Ostpreußen. Der Deutsche Forstwirt 16:845-847; 853-856.
202. Holmes, F.N. 1961. Salt injury to trees. Phytopathology 51:712-718.
203. Holt, R.A. 1988. The Maui Forest Trouble: reassessment of an historic forest dieback. M.S. Thesis. Univ. of Hawaii. Honolulu, HI.
204. Horner, R.M. 1956. A Diaporthe canker of Betula lutea. Proc. Can. Phytopathol. Soc. 23:16-17.
205. Houston, D.R. 1967. The dieback and decline of northeastern hardwoods. Trees 28:12-14.
206. Houston, D.R. 1971. Noninfectious diseases of oak. Pages 118-123 in: Oak Symposium Proceedings; 1971 Aug. 16-20; Morgantown, WV. USDA Northeastern Forest Experiment Station.
207. Houston, D.R. 1973. Diebacks and declines: diseases initiated by stress, including defoliation. Proc. Int. Shade Tree Conf. 49:73-76.
208. Houston, D.R. 1975. Beech bark disease -- the aftermath forests re structured for a new outbreak. J. For. 73:660-663.
209. Houston, D.R. 1981. Stress triggered tree diseases - the diebacks and declines. USDA. For. Serv. NE-INF-41-81. 36 pp.
210. Houston, D.R. 1984. Stress related to diseases. Arbor. J. 8:137-149.
211. Houston, D.R. 1985. Dieback and declines of urban trees. J. Arbor. 11(3):65-72.
212. Houston, D.R. 1986. Insects and diseases of northern hardwood ecosystems. Pages 109-138 in: Proc. Conf. on the Northern Hardwood Resource Management and Potential. G.D. Mroz and D.D. Reed eds. Aug., 1986. Mich. Tech. Univ., Houghton. 434 pp.
213. Houston, D.R. 1987. Forest tree declines of past and

present: current understanding. Can. J. Plant Pathol. 9:349-360.
214. Houston, D.R. 1992. A host-stress-saprogen model for forest dieback-decline diseases. (Chapter 1).
215. Houston, D.R., Allen, D.C., and Lachance, D. 1990. Sugarbush management: A guide to maintaining tree health. USDA For. Serv. Gen. Tech. Rep. NE-129. 55 pp.
216. Houston, D.R., and Kuntz, J.E. 1964. Pathogens associated with maple blight. Pages 58-79 in: Part III. Studies of maple blight. Madison, University Wisconsin Agric. Res. Bull. 250.
217. Houston, D.R., Mahoney, E.M., and McGauley, B.H. 1987. Beech bark disease: association of Nectria ochroleuca in W. VA., PA., and Ontario. Phytopathology 77:1615.
218. Houston, D.R., Parker, E.J., Perrin, R., and Lang, K.L. 1979. Beech bark disease: a comparison of the disease in North America, Great Britain, France and Germany. Eur. J. For. Path. 9:199-211.
219. Hoyle, M.C. 1965. Growth of yellow birch in a podzol soil. USDA Forest Serv. Res. Pap. NE-38. 14 pp.
220. Hoyle, M.C. 1969. Response of yellow birch in acid subsoil to macronutrient additions. Soil Sci. 103:354-358.
221. Hursch, C.R., and Haasis, F.W. 1931. Effects of 1925 summer drought on souther Appalachian hardwoods. Ecology 12:380-386.
222. Innes, J.L. 1988. Forest health surveys -- a critique. Env. Poll. 54:1-15.
223. Innes, J.L. and Boswell, R.C. 1990. Reliability, presentation, and relationships among data from inventories of forest conditions. Can. J. For. Res. 20:790-799.
224. Isaacs, E.H., and Srivastava, R.M. 1989. Applied geostatistics. Oxford University Press, New York.
225. Itow, S., and Mueller-Dombois, D. 1988. Population structure, stand-level dieback and recovery of Scalesia pedunculata forest in the Galapagos Islands. Ecol. Res. 3:333-339.
226. Jacobi, J.D., Gerrish, G., Mueller-Dombois, D., and Whiteaker, L. 1988. Stand-level dieback and Metrosideros regeneration in the montane rainforest of

Hawaii. GeoJ. 17:193-200.
227. Jakucs, P. 1988. Ecological Approach to Forest Decay in Hungary. Ambio 17(4):267-274.
228. Johnson, A.H., Cook, E.R., and Siccama, T.G. 1988. Climate and red spruce growth and decline in the northern Appalachians. Proc. Nat. Acad. Sci. 85:5369-5373.
229. Johnson, A.H., Friedland, A.J., and Dushoff, J.C. 1986. Recent and historic red spruce mortality: Evidence of climatic influence. J. Water Air Soil Pollut. 30:319-330.
230. Johnson, A.H., and Siccama, T.J. 1983. Acid deposition and forest decline. Environ. Sci. Technol. 17:249-304.
231. Johnson, W.T., and Lyon, H.H. 1976. Insects that feed on trees and shrubs. Cornell Univ. Press. Ithaca, NY. 464 pp.
232. Journel, A.G., and Huijbregts, C.H. 1978. Mining Geostatistics. Academic Press, London.
233. Kandler, O. 1985. Imissions-versus Epidemie-Hypothesen. Pages 20-50 in: Waldschäden. G. von Kortzfleisch, ed. Oldenbourg Verlag, München, FRG.
234. Kandler, O. 1988. Epidemiologische Bewertung der Waldschadenserhebungen 1983 bis 1987 in der Bundesrepublik Deutschland. Allg. Forst Jagdztg. 159:179-194.
235. Kandler, O. 1988. Lichen and conifer recolonization in Munich's cleaner air. Pages 784-790 in: Air Pollution and Ecosystems. Proceedings of an Intern. Symp., Grenoble, France, 18-22 May 1987. P. Mathy, ed. D. Reidel Dordrecht Boston, Tokyo.
236. Kandler, O.E. 1990. Epidemiological evaluation of the development of Waldsterben in Germany. Plant Disease 74:4-12.
237. Kandler, O. 1991. Verlauf des Tannensterbens in Ostbayern und den Bagerischen Alpen. Proc.
238. Kandler, O. 1992. Historical declines and diebacks in central European forests and present conditions. Environ. Toxicology and Chem. 10:
239. Kandler, O., and Miller, W. 1991. Dynamics of "acute yellowing" in spruce connected with Mg deficiency. Water Air Soil Pollut. 54:21-34.
240. Kandler, O., Miller, W., and Ostner. 1987. Dynamik der "akuten Vergilbung" der Fichte. Epidemiologische

und physiologische Befunde. Allg. Forst. Z. 42:715-723.
241. Kandler, O., Senser, M., and Miller, W. 1990. Vergilbung und Widerergrünung der Fichten. Pages 113-137 in: Neuartige Waldschäden - Erkenntnisse und Folgerungen. J. Jositz, ed. vol. 56, Berichte & Studien der Hanns-Seidel-Stiftung e.V., München, ISBN 3-88795-076-3.
242. Karnig, J.J., and Lyford, W.H. 1968. Oak mortality and drought in the Hudson Highlands. Harvard Black Rock For. Pap. No. 29. Cornwall, NY. 13 pp.
243. Kaufmann, W. 1989. Air pollution and forests: An update. Amer. For. 95(5,6):37-44.
244. Ke, J., and J.M. Skelly. 1990. Foliar symptoms on Norway spruce and relationships to magnesium deficiency. J. Water, Air, Soil Pollution (In press).
245. Kegg, J.D. 1973. Oak mortality caused by repeated gypsy moth defoliations in New Jersey. J. Econ. Ent. 66:639-641.
246. Kelley, R.S. 1988. The relationship of defoliators to recent hardwood dieback and decline in Vermont. Page 47 in: Proc. 21st. Ann. NE For. Insect Work Conference. A.G. Raske (compiler). Albany, NY.
247. Kelly, J.M., Schaedle, M., Thornton, F.C., and Joslin, J.D. 1990. Sensitivity of tree seedlings to aluminum: II. Red oak, sugar maple, and European beech. J. Env. Qual. 19:172-179.
248. Kennel, E. 1989. Entwicklung und Stand der waldschadenserhebung. Pages 43-59 in: Proc. Inter. Congress on For. Decline Research: State of knowledge and perspectives. Ulrich, B., ed. Kernforschungszentrum Karlsruhe GmbH. Postfach 3640, 7500 Karlsruhe 1, FRG.
249. Kennel, E., and Reitter, A. 1989. Waldschadensinventur Bayern: Ergebnisse 1986-1988. Schriftenreihe Forstwiss. Fakultät Univ. München Nr. 94.
250. Kerr, R.A. 1989. Hansen vs. the world on the greenhouse threat. Science 244:1041-1043.
251. Kessler, K.J., Jr. 1963. Dieback of sugar maple, Upper Michigan - 1962. USDA For. Serv. Res. Note LS-13. 2 pp.
252. Kessler, K.J., Jr. 1978. How to control sapstreak disease of sugar maple. USDA For. Serv., North Central For. Exp. Stn. 5 pp.

253. Kiester, E., Jr. 1985. A deathly spell is hovering above the Black Forest. Smithsonian. 20(8):1-12.
254. Klapp, E., and R. Spennemann. 1933. ökologie und Abbau der Kartoffel. II. Klima, Boden und Wildflora in gesunden abbauenden Lagen. Pflanzenbau 9:303-313.
255. Klecka, W.R. 1980. Discriminant analysis. Sage University Paper series on Quantitative Applications in the Social Sciences, 07-019. Beverly Hills and London.
256. Klein, R.M., and Perkins, T.D. 1988. Primary and secondary causes and consequences of contemporary forest decline. Bot. Rev. 54:2-42.
257. Klinger, L.F. 1988. Successional change in vegetation and soils of southeast Alaska. Ph.D. Thesis. University of Colorado, Boulder, CO. 234 pp.
258. Knox, J.L. 1981. Atmospheric blocking in the northern hemisphere. Internal Rpt. 82-2 Canadian Climate Center, Atmospheric Environment Service, Environment Canada Downsview, Ontario, Canada. 244 pp.
259. Knull, J.N. 1932. Observations of three important forest insects. J. Econ. Ent. 25:1196-1202.
260. Koch, H. 1978. Erhebung über Schäden an der Tanne in Ostbayern. Allg. Forest. Z. 23:989-991.
261. Kohyama, T. 1988. Etiology of "Shimagare" dieback and regeneration in subalpine Abies forests of Japan. GeoJournal 17:201-208.
262. Killar, A., Seemuller, E., Bonnet, F., Saillard, C., and Bové, J.M. 1990. Isolation of the DNA of various plant pathogenic mycoplasma-like organisms from infected plants. Phytopathology 80:233-237.
263. Krahl-Urban, B., Papke, H.E., Peters, K., and Schimansky, C. 1988. Forest Decline: Cause-effect research in the United States of North America and the Federal Republic of Germany. Assess. Group for Biol. Ecol. and Energy, Julich Nuclear Res. Center, USEPA, CERL., Corvallis, OR. 137 pp.
264. Krause, G.H.M., Arndt, U., Brandt, C.J., Bucher, J., Kenk, J., and Matzner, E. 1986. Forest decline in Europe: development and possible causes. J. Water, Air and Soil Poll. 31:647-668.
265. Krause, G.H.M., und Prinz, B. 1989. Experimentelle Untersuchungen der LIS zur Aufklärung möglicher Ursachen der neuartigen Waldschäden. LIS-Berichte Nr. 80. Landesanstalt für Immissionsschutz Nordhein-

Westfalen, Essen.
266. Lacasse, N.L., and Rich, A.E. 1964. Maple decline in New Hampshire. Phytopathology 54:1071-1075.
267. Lachance, D. 1985. Répartition géographique et intensité du dépérissement de l'érable à sucre dans les érablières au Québec. Phytoprotection 66:83-90.
268. Lachance, D. 1987. Commentary in: L. Chartrand, Qui tue les arbres? L'actualité (December(39-44).
269. Lana, A.O., and Agrios, G.N. 1974. Transmission of a mosaic disease of white ash to woody and herbaceous hosts. Plant Dis. Rep. 58:536-540.
270. Landmann, G. 1989. Comments on the current state of knowledge and further needs of research into forest decline. Pages 946-955 in: Proceedings International Congress on Forest Decline Research: State of knowledge and perspectives. Ulrich, B., ed. Kernforschungszentrum Karlsruhe GmbH. Postfach 3640, 7500 Karlsruhe 1.
271. Larcher, W., and Bauer, H. 1981. Ecological significance of resistance to low temperature. Pages 403-437 in: Physiological Plant Ecology. O.L. Lange, P.S. Nobel, C.B. Osmond and H. Ziegler eds. Volume 1, Springer Verlag, New York.
272. Law, J.R., and Gott, J.D. 1987. Oak mortality in the Missouri Ozarks. Pages 427-436 in: Proc. 6th Central Hardwood For. Conf. Hay, R.L., F.W. Woods, H. DeSelm, eds. 24-26 February 1987; Knoxville, TN. Univ. Tennessee.
273. Leaphart, C.D., and Stage, A.R. 1971. Climate: a factor in the origin of the pole blight disease of Pinus monticola Dougl. Ecology 52:229-239.
274. LeBlanc, D.C. 1990. Red spruce decline on Whiteface Mountain, New York. I. Relationships with elevation, tree age, and competition. Can. J. For. Res. 20:1408-1414.
275. LeBlanc, D.C., and Raynal, D.J. 1990. Red spruce decline on Whiteface Mountain, new York. II. Relationships between apical and radial growth decline. Can. J. For. Res. 20:1415-1421.
276. Legendre, P., and Fortin, J.J. 1989. Spatial pattern and ecological analysis. Vegetation 82:107-138.
277. Lewis, R., Jr. 1981. Hypoxylon spp., Ganoderma lucidum and Agrilus bilineatus in association with

drought related oak mortality in the South. Phytopathology 71:890.
278. Liming, F.G. 1957. Home made dendrometers. J. For. 55:575-577.
279. Lohman, M.L., and Watson, A.J. 1943. Identity and host relations of Nectria species associated with diseases of hardwoods in the eastern States. Lloydia 6:77-108.
280. Long, W.H. 1914. The death of chestnuts and oaks due to Armillaria mellea. USDA Bull. 89. 9 pp.
281. Lortie, M., and Pomerleau, R. 1964. Fluctuations du dépérissement du bouleau à papier dans le Québec, de 1953 à 1961. Phytoprotection 45:77-82.
282. Lucaschewski, Von I., and Mettendorf, B. 1988. Waldschadens-situation 1988 in Baden-Wurttemberg. Forst und Holz, Nr. 20:506-510.
283. Lucier, A.A., and B.B. Stout. 1988. Changes in forests and their relationship to air quality. TAPPIJ. 70:103-107.
284. Ma, G. 1989. Forest decline and its possible causes in south China. (Abstr.) Poster Paper 394 in Proc. Int'l. Cong. on Forest Decline Research: State of Knowledge and Perspectives. Lake Constance, FRG, Fed. Res. Advis. Board on Forest Decline/Air Poll. of the Fed. and State Gov't., Karlsruhe, FRG.
285. MacKenzie, J.J. 1988. Breathing easier: taking action on climate change, air pollution, and energy insecurity. World Resources Institute. Washington, DC. 23 pp.
286. MacKenzie, J.J., and El-Ashry, M.T. 1988. III Winds: Air pollution's toll on trees and crops. World Resources Institute, Washington, DC. 74 pp.
287. Maddixo, J. 1972. The doomsday syndrome. Sat. Rev. (Oct. 21):31-37.
288. Magasi, L.P. 1986. Forest pest condition in the Maritimes 1985. Information Report M-X-161. Maritimes Forestry Centre, Canadian Forestry Service, Fredericton, New Brunswick, Canada.
289. Magnuson, J.J. 1990. Long-term ecological research and the invisible present. BioScience 40:495-501.
290. Malek, J. 1981. Problematik der ökologie der Tanne (Abies alba Mill.) und ihres Sterbens in der CSSR. Forstwiss. Centralb. 100:170-174.
291. Manion, P.D. 1981. Tree disease concepts. Prentice Hall, Englewood Cliffs, NJ. 399 pp.

292. Manion, P.D. 1985. Effects of air pollution on forest: Critical review prepared discussion. J. Air Pollut. Control Assoc. 35:919-922.
293. Manion, P.D. 1987. Decline as a phenomenon in forests: Pathological and ecological considerations. Pages 267-275 in: Proc. NATO Advanced Workshop. The Effects of Atmospheric Pollutants on Forests, Wetlands and Agricultural Ecosystems. T.C. Hutchinson and K.M. Meema, eds. Springer-Verlag, Berlin.
294. Manion, P.D. 1989. Hardwood forest decline - concepts and management. Pages 127-130 in: Proc. Soc. Amer. For. Convention, Oct. 16-19, 1988. Rochester, NY.
295. Manion, P.D. 1991. Tree disease concepts. Prentice Hall, Englewood Cliffs, NJ. 409 pp.
296. Maramorosch, K., and Rayschaudhuri, S.P. eds. 1981. Mycoplasma diseases of trees and shrubs. Academic Press, New York.
297. Marshall, R.P. 1930. A canker of ash. Pages 128-130 in: Proceedings Sixth Annual Meeting National Shade Tree Conference.
298. Martin, J.R. 1989. Vegetation and environment in old growth forests of northern southeast Alaska: a plant association classification. MS Thesis. Arizona State Univ., Tempe, AZ. 221 pp.
299. Mathysse, A.G., and Scott, T.K. 1984. Functions of hormones at the whole plant level of organization. Pages 219-243 in: Encyclopedia of Plant Physiology, Vol. 10. T.K. Scott, ed. Springer-Verlag, Berlin.
300. Matteoni, J.A. 1983. Yellows diseases of elm and ash: water relations, vectors and role of yellows in ash decline. Ph.D. Thesis. Cornell University. Ithaca, NY. 152 pp.
301. Matteoni, J.A., and Sinclair, W.A. 1983. Stomatal closure in plants affected with mycoplasma-like organisms. Phytopathology 73:396-402.
302. Mattson, K.G., Arnaut, L.Y., Reams, G.A., Cline, S.P., Peterson, C.E., and Vong, R.J. 1990. Response of forest trees to sulfur, nitrogen, and associated pollutants. U.S. Env. Prot. Agency Rep. EPA/600/3-90/074. 134 pp.
303. Mayer, H. 1979. Zur waldbaulichen Bedeutung der Tanne im mitteleuropäischen Bergwald. Forst-u. Holzwirt 34:333-343.

304. Mayer, H., König, C., and Rall, A. 1988. Identification of meteorological events that cause plant-physiological stress in forest trees. Forstw. Cbl. 107:131-140.
305. Mayer, P. 1989. Wenn die baume schreien konnten. Stern Magazine. 45:29-60.
306. Maynard, C.A., Overton, R.P., and Johnson, L.C. 1987. The silviculturist's role in tree improvement in northern hardwoods. Pages 35-45 in: Managing Northern Hardwoods -Proceedings of a Silvicultural Symposium. R.D. Nyland ed. June 1986. SUNY Coll. Environ. Sci. & For., Syracuse, NY. Fac. For. Misc. Publ. No. 13 SAF Pub. 87-03.
307. McCreery, L.R., Miller-Weeks, M., Weiss, M.I., and Millers, J. 1986. Cooperative survey of red spruce and balsam fir in New York, Vermont, and New Hampshire: A progress report. Pages 167-180 in: Proceedings Integrated Pest Management Symposium 24-27 March 1986, Madison, WI.
308. McIlveen, W.D., Rutherford, S.T., and Linzon, S.N. 1 986. A historical perspective of sugar maple decline within Ontario and outside of Ontario. Ont. Min. Environ. Rept. No. ARB-141-86-Phyto. 40 pp.
309. McIntyre, A.C., and Schnur, G.L. 1936. Effects of drought on oak forests. State College, PA, Agric. Exp. Sta. Bull. 325. 43 pp.
310. McLaughlin, D.L. 1987. Canada Department of the Environment 1987. Forest Decline: Report on the workshop at Wakefield, Quebec, October 20-22, 1986. Atmospheric Environment Service, Federal LRTAP Liaison Office, Downsview, Ontario, Canada.
311. McLaughlin, D.L., and Butler, J. 1987. Summary Report of Questionnaires Distributed to the Ontario Maple Syrup Producers Association - Fall 1985. Ontario Ministry of the Environment Rept. No. ARB-142-86-PHYTO.
312. McLaughlin, D.L., Corrigan, D.E., and McIlveen, W.D. 1992. Etiology of Sugar Maple Decline at Selected Sites in Ontario (1984-1990). Ontario Ministry of the Environment Report No. ARB-052-92-PHYTO.
313. McLaughlin, D.L., Linzon, S.N., Dimma, D.E., and McIlveen, W.D. 1985. Sugar maple decline in Canada. Ontario Min. of Environment Rept. No. ARB-144-85-Photo. 18 pp.

314. McLaughlin, S.B. 1985. Effects of air pollution on forests: a critical review. J. Air Poll. Control Assoc. 35:512-534.
315. Medwedjew, S. 1971. Der Fall Lyssenko. Hoffmann and Campe Hamburg, FRG.
316. Mehr, C. 1989. Are the Swiss forests in peril? Nat'l. Geog. 175:637-651.
317. Mielke, M.E., Soctomah, D.G., Marsden, M.A., and Ciesla, W.M. 1986. Decline and mortality of red spruce in West Virginia. USDA For. Serv. Forest Pest Mgmt. Methods App. Group Rep. No. 86-4. 26 pp.
318. Miller, P.R., and O'Brien, M.J. 1951. An apparently new sweetgum disease in Maryland. Plant Dis. Rep. 35:295-297.
319. Millers, I., Shriner, D.S, and Rizzo, D. 1989. History of hardwood decline in the eastern United Sates. USDA For. Serv. Tech. Rep. NE-126. 75 pp.
320. Moomaw, J.C., Nakamura, M.T., and Sherman, G.D. 1959. Aluminum in some Hawaiian plants. Pac. Sci. 8:335-341.
321. Moore, H.R., Anderson, W.R., and Baker, R.H. 1951. Ohio maple syrup - some factors influencing production. Ohio Agric. Exp. Stn. Res. Bull. 718. 53 pp.
322. Morrow, R.R. 1955. Influence of tree crowns on maple sap production. Cornell Agric. Exp. Stn. Bull. 916. 30 pp.
323. Morstatt, H. 1925. Entartung, Altersschwäche und Abbau bei Kulturpflanzen, insbesondere der Kartoffel. Naturwissenschaft und Landwirtschaft Heft 7, 1-74. Verlag Dr. F.P. Datterer und Cie., Fresiing-München, FRG.
324. Mosher, D.G., and Simmons, G.A. 1978. Michigan pest conditions - 1977. Michigan Cooperative Forest Pest Management Program. Lansing, MI. 21 pp.
325. Mueller-Dombois, D. 1983. Canopy dieback and successional processes in Pacific forests. Pac. Sci. 37:317-325.
326. Mueller, Dombois, D. 1985. 'Ohi'a dieback in Hawaii: 1984 synthesis and evaluation. Pac. Sci. 39:150-170.
327. Mueller-Dombois, D. 1986. Perspectives for an etiology of stand-level dieback. Ann. Rev. Ecol. Syst. 17:221-243.
328. Mueller, Dombois, D. 1987. Natural dieback in forests.

BioScience 37:575-583.
329. Mueller-Dombois, D. 1988. Vegetation dynamics and slope management on the mountains of the Hawaiian Islands. Env. Conserv. 15:255-260.
330. Mueller-Dombois, D. 1988. Forest decline and dieback - a global ecological problem. Trends Ecol. Evolut. 3:310-312.
331. Mueller-Dombois, D. 1989. Perspective for an etiology of stand level dieback. Ann. Rev. Ecol. Syst. 17:221-243.
332. Mueller-Dombois, D., Canfield, J.E., Halt, R.A., and Buelow, G.P. 1983. Tree-group death in North America and Hawaiian forest: A pathological problem or a new problem for vegetation ecology? Phytoenologia 11:117-137.
333. Mueller-Dombois, D., Jacobi, J.D., Cooray, R.G., and Balakrishnan, N. 1980. 'Ohi'a rain forest study: ecological investigations of the 'ohi'a dieback problem in Hawaii. College of Trop. Ag. & Human Resources. Hawaii Ag. Expt. Sta. Miscell. Public. 183, 64 pp.
334. Mueller-Dombois, D., and McQueen, D.R. eds. 1983. Canopy dieback and dynamic processes in Pacific forests. Pac. Sci. 37(4), Special sympos. issue with 17 papers.
335. Mülder, D. und Zycha, H. 1980. Saubere Wirtschaft im Buchenrevier. Ein Weg zue Eindämmung des Buchensterbens? Holz-Zentralbl. 63:266-269.
336. Münch, E. 1924. Die künftige Leistungsfähigkeit der deutschen Forstwirtschaft vom Standpunkt der Biologie betrachtet. Tharandter Forstl. Jahrb. 75:1-27.
337. Neger, F.W. 1908. Das Tannensterben in den sächsischen und anderen deutschen Mittelgebirgen. Tharandter Forstl. Jahrb. 58:201-225.
338. Neilsen, R.P., King, G.A., DeVelice, R.L., Lenihan, J., Marks, D., Dolph, J., Campbell, B., and Glick, B. 1989. Sensitivity of ecological landscapes and regions to global climatic change. United States Environmental Protection Agency. Corvallis, OR.
339. Nicholas, N.S., and Zedaker, S.M. 1989. Ice damage in spruce-fir forests of the Black Mountains, North Carolina. Can. J. For. Res. 19:1487-1491.
340. Nicholas, N.S., and Zedaker, S.M. 1990. Forest decline and regeneration success in the Great Smoky Mountains

spruce-fir. Page 44 in: Proc. First Ann. Appalachian Man and the Biosphere Conf., Nov. 4-5, Gatlinburg, TN. TVA/LR/NRM-90/8.
341. Nienhaus, F., and Castello, J.D. 1989. Viruses in forest trees. Annu. Rev. Phytopathol. 27:165-186.
342. Niesslein, E. 1984. Zu den Folgen des Waldsterbens. Holz-Zentralbl. 67:1-16.
343. Nihlgard, B. 1990. Relationship of forest damage to air pollution in Nordic countries. Agr. For. Meteorol. 50:87-89.
344. Nyland, R.D. 1986. Important trends and regional differences in silvicultural practice for northern hardwoods. Pages 156-182 in: Proc. of a Conf. on the Northern Hardwood Resource: Management and potential. G.D. Morz and D.D. Reed (compilers). Aug. 1986. Mich. Tech. Univ., Houghton, MT.
345. Ogden, J. 1988. Forest dynamics and stand-level dieback in New Zealand's Nothofagus forests. GeoJ. 17:225-230.
346. Olkesyn, J., and Przybyl, K. 1987. Oak decline in the Soviet Union - Scale and hypotheses. Eur. J. For. Path 17:321-336.
347. Pare, D., and Bernier, B. 1989. Origin of the phosphorus deficiency observed in declining sugar maple stands in the Quebec Appalachians. Can. J. For. Res. 19:24-34.
348. Parker, J. 1970. Effects of defoliation and drought on root food reserves in sugar maple seedlings. USDA For. Serv. Res. Paper NE169. 4 pp.
349. Parker, J. 1974. Effects of defoliation, girdling, and severing of sugar maple trees on root starch and sugar levels. USDA For. Serv. Res. Pap. NE-306. 4 pp.
350. Parker, J., and Houston, D.R. 1971. Effects of repeated defoliation on root and root collar extractives of sugar maple trees. For. Sci. 17:91-95.
351. Parker, J., and Patton, R.L. 1975. Effects of drought and defoliation on some metabolites in roots of black oak seedlings. Can. J. For. Res. 5:457-463.
352. Payette, S., Filion, L., Gauthier, L., and Boutin, Y. 1985. Secular climate change in old-growth tree-line vegetation of northern Quebec. Nature 315:135-138.
353. Pearce, F. 1985. The strange death of Europe's trees. New Sci. 4:41-45.

354. Peterman, R.M. 1990. The importance of reporting statistical power: the forest decline and acid deposition example. Ecology 71:2024-2027.
355. Pickett, S.T.A., and McDonnell, M.J. 1989. Changing perspectives in community dynamics: a theory of successional forces. Tree 4:241-245.
356. Pickett, S.T.A., and White, P.S. ed. 1985. The ecology of natural disturbance and patch dynamics. Academic Press, Inc., New York. 472 pp.
357. Pitelka, L.F., and Raynal, D.J. 1989. Forest decline and acid deposition. Ecology 79:2-10.
358. Platt, W.J., and Strong, D.R. 1989. Special feature: gaps in forest ecology. Ecology 70:535.
359. Podzer, F.D. 1980. Definition and diagnosis of eucalyp diebacks. in: Eucalypt Dieback in Forest and Woodland. K.M. Old, G.A. Kile and Ohmart, P. eds. Proceedings of Conference, CSIRO Division of Forest Research, 6-4 August 1980, Canberra, Australia.
360. Pomerleau, R. 1944. Observation sur quelques maladies non-parasitaires des arbres dans le Québec. Can. J. Res. (Section C) 22:171-189.
361. Pomerleau, R. 1953. History of hardwood species dying in Quebec. Pages 10-11 in: report of the Symposium on birch dieback. Canada Department of Agriculture, Ottawa, Ontario, Canada.
362. Pomerleau, R. 1953. The relationship between environmental conditions and the dying of birches and other hardwood trees. Pages 114-117 in: Report of the symposium on birch dieback. 21-22 March 1953. Canadian Department Agriculture. Ottawa, Ontario, Canada.
363. Pomerleau, R. 1991. Experiments on the causal mechanisms of dieback on deciduous forests in Quebec. Forestry Canada, Quebec Region. Information Report LAU-X-96, Sainte Foy, Quebec, Canada.
364. Prinze, B. 1987. Causes of forest damage in Europe: Major hypothesis and factors. Environ. 29(9):10-19.
365. Prinz, B., Krause, G.H.M., and Jung, K.D. 1985. Untersuchungen der LIS zur Problematik der Wäldschaden Pages 143-194 in: Wäldschaden Kortzfleisch, G. von, ed.) Oldenbourg, Verlag, München, FRG.
366. Prinz, B., Krause, G.H.M. and Stratmann, H. 1982.

Waldschäden in der Bundesrepublik Deutschland. LIS-Berichte 28, Landesanstalt für Immissionsschutz Nordrhein-Westfalen, Essen, FRG.

367. Pusey, P.L. 1989. Influence of water stress on susceptibility on nonwounded peach bark to Botryosphaeria dothidea. Plant Dis. 73(12):1000-1003.

368. Quimby, J.W. 1985. Tree mortality in Pennsylvania forests defoliated by the gypsy moth -- a 1984 update. Department Environmental Resources. Bureau Forestry, Division Forest Pest Management. Middletown, PA. 11 pp.

369. Raymond, F.L., and Reid, J. 1961. Dieback of balsam fir in Ontario. Can. J. Bot. 392:233-251.

370. Rebel, K. 1920. Streunutzung, insbesondere im bayerischen Staatswald. Verlag Jos. E. Huber, Diessen von München, 172 pp.

371. Rebel, K. 1920. Waldbauliches aus Bayern. II. Band, Verlag Jos. E. Huber, Diessen von München, 228 pp.

372. Redmond, D.R. 1955. Rootlets, mycorrhiza, and soil temperatures in relation to birch dieback. Can. J. Bot. 33:595-672.

373. Rehfuess, K.E. 1986. On the causes of decline of Norway spruce (Picea/Abies Krast.) in Central Europe. Soil Use Manage. 1:30-31.

374. Rehfuess, K.E. 1987. Perceptions on forest diseases in central Europe. Forestry 60:1-11.

375. Reich, R.W. 1990. Causes of dieback of douglas-fir in the interior of British Columbia. M.S. Thesis. Department of Forestry, University of British Columbia, Vancouver, British Columbia. Canada. 123 pp.

376. Rippel, K. 1948. Zur Frage des Kartoffelabbaues. Zeitschr. Pflanzenkrankheiten 55:351.

377. Rizzo, D.M., and Harrington, T.C. 1988. Root movement and wind damage of red spruce and balsam fir on subalpinc sites int eh White Mountains, New Hampshire. Can. J. For. Res. 18:991-1001.

378. Rizzo, D.M., and Harrington, T.C. 1988. Root and butt rot fungi on red spruce and balsam fir in the White Mountains, New Hampshire. Plant Dis. 72:329-331.

379. Roberts, T.M., Skeffington, R.A., and Blank, L.W. 1989. Cause of Type I spruce decline in Europe. Forestry 62:180-222.

380. Roloff, A. 1989. Morphological changes in the crowns

of European beech (Fagus sylvatica L.) and other deciduous tree species. Pages 81-107 in: Proc. Inter. Congress on Forest Decline Research: State of knowledge and perspectives. Ulrich, B., ed. Kernforschungszentrum Karlsruhe GmbH. PF 3640, 7500 Karlsruhe 1.

381. Ross, E.W. 1966. Ash dieback. Etiological and developmental studies. State University College of Forestry at Syracuse Tech. Pub. 88, Syracuse, NY.
382. Rowe, R.S. 1972. Forest Regions of Canada. Forestry Canada, Pub. No. 1300.
383. Rubner, K. 1936. Beitrag zur Kenntnis der Fichtenformen und Fichtenrassen. Tharandter Forstl. Jahrb. 87:101-176.
384. Rubner, K. 1939. Beitrag zur Kenntnis der Fichtenformen und Fichtenrassen. Tharandter Forstl. Jahrb. 90:883-915.
385. Ruge, U. 1950. Über die möglichen Ursachen des Buchensterbens. Allg. Forst Z. 5:217-219.
386. Rush, P.A. 1986. Forest pest conditions report for the northeastern area -- 1985. USDA For. Serv. State and Private For. Pub. No. NA-FR-33. 35 pp.
387. Russell, K.W. 1965. Conifer freeze damage 1964-65, Snohmish County, Washington, Resource Management Report No. 11. Washington Department of Natural Resources, Olympia, WA.
388. Ruth, R.H., and Harris, A.S. 1979. Management of western hemlock-Sitka spruce forests for timber production. USDA For. Serv. Gen. Tech. Rep. PNW-88. 197 pp.
389. Sakai, A., and Larcher, W. 1987. Frost survival of plants, Ecological Studies 62. Springer-Verlag, Berlin.
390. Sanderson, M.E., and Phillips, D.W. 1967. Average annual water surplus in Canada. Climatological Studies No. 9, Meteorological Branch, Canada Department of Transport, Toronto, Ontario, Canada.
391. Sargent, R.H., and Moffit, F.H. 1929. Aerial photographic surveys in southeastern Alaska. U.S. Geological Survey Bulletin 797-E. GPO.:143-160.
392. Saxena, V.K., Stogner, R.E., Hendler, A.H., DeFelice, T.P., Yeh, R.J.Y., and Lin, N.H. 1989. Monitoring the chemical climate of Mt. Mitchell State Park for evaluation of its impact on forest decline. Tellus

41B:92-109.
393. Schall, R.A., and Agrios, G.N. 1973. Graft transmission of ash witches'-broom to ash. Phytopathology 63:206.
394. Schlaepfer, R., und Hämmerli, F. 1990. Das "Waldsterben" in der Schweiz aus heutiger Sicht. Schweiz. Z. Forstwes. 141:163-188.
395. Schlenker, G. 1976. Einflüsse des Standorts und der Bestandsverhältnisse auf die Rotfäule (Kernfäule) der Fichte. Beih. Forstwiss. Centralbl. 36:47-57.
396. Schmidt, R.A., and Fergus, C.L. 1965. Branch canker and dieback of Quercus prunis caused by a species of Brotryodiplodia. Can. J. Bot. 43:731-737.
397. Schneider, S.H. 1989. The greenhouse effect: science and policy. Science 243:771-781.
398. Schneider, T.W., Lorenz, M., und Poker, J. 1987. Abschätzung der erträglichen Folgen der neuartigen Waldschäden im Bereich der Landesforstverwaltung Hamburg mit Hilfe dynamischer Wachstumsmodelle. Mitteilungen der Bundesforschungsanstalt für Forst-jund Holzwirtschaft Hamburg FRG. 155:61-77.
399. Schoeneweiss, D.F. 1978. The influence of stress on diseases of nursery and landscape plants. J. Arbor. 4:217-225.
400. Schoeneweiss, D.F. 1981. Infectious diseases of trees associated with water and freezing stress. J. Arbor. 7:13-18.
401. Schoeneweiss, D.F. 1981. The role of environmental stress in diseases of woody plants. Plant Dis. 65:308-314.
402. Schopfer, W., and J. Hradetzky. 1984. Circumstantial evidence: Air pollution is the determinative factor causing forest decline. Forst-Wissenschaft. Centralblatt 103:231-247.
403. Schröter, H. 1983. Krankheitsentwicklung von Tannen und Fichten auf Beobachtungsflächen der FVA in Baden-Württemberg. Allg. Forst Z. 38:648-649.
404. Schuh, Van H. 1989. Die Waldsterben da Waldmuchern. Die Zeit 80 Wissenschaft, November 16 edition.
405. Schulze, E.D. 1989. Air pollution and forest decline in a spruce (Picea abies) forest. Science 224:776-783.
406. Schulze, E.D., Lange, O.L., and Oren, R. 1989. Forest decline and air pollution. Springer-Verlag, Heidelberg.

Ecol. Studies 77, 475 pp.
407. Schütt, P. 1982. Aktuelle Schäden am Wald-Versuch einer Bestandsaufnahme. Holz-Zentralbl. 25:369-372.
408. Schütt, P. 1984. Der Wald stirbt an Stress. C. Bertelsmann Verlag GmbH., München, FRG.
409. Schütt, P. 1989. Forest decline in Germany, Pages 87-88 in Proc. US/FRG Research Symposium: Effects of Atmospheric Pollutants on the Spruce-fir Forests of the Eastern United States and the Federal Republic of Germany. USDA-For. Serv. Gen'l. Tech. Rept. NE-120. 543 pp.
410. Schütt, P., and Cowling, E.B. 1985. Waldsterben, a general decline of forests in Central Europe: Symptoms development, and possible causes. Plant Dis. 69:548-558.
411. Schütt, P., Koch, W., Blaschke, H., Lang, K.J., Reigber, E., Schuck, H.J., and Summerer, H. 1985. So stirbt der Wald. BLV-Verlag, München, FRG. 127 pp.
412. Schütt, P., Koch, W., Blaschke, H., Lang, K.J., Schuck, H.J. und Summerer, H. 1983. So stirbt der Wald. BLV Verlag, München, FRG. 95 pp.
413. Schweingruber, F.H. 1989. Bäume schweizerischer Gebirgswälder auf alten und neuen Postkarten. Allg. Forst Z. 44:262-268.
414. Schweingruber, F.H., Kontic, R., and Winkler-Seifert, A. 1983. Eine jahrringanalytische Studie zum Nadelbaumsterben in der Schweiz. Eidgenössische Anstalt für das forstliche Versuchswesen, CH-Birmensdorf, Berichte Nr. 253:1-29.
415. Scott, J.T., Siccama, T.G., Johnson, A.H., and Breisch, A.R. 1984. Decline of Red Spruce in the Adirondacks, New York. Bulletin of the Torrey Botanical Club. 111:438-444.
416. Seemüller, E. 1989. Mycoplasmas as the cause of diseases in woody plants in Europe. Forum Mikrobiol. 12:144-151.
417. Seemüller, E., and W. Lederer, W. 1988. MLO-associated decline of Alnus glutinosa. Populus tremula and Crataegus monogyna. J. Phytopathol. 121:33-39.
418. Shaw, C.G. III, Eglitis, A., Laurent, T.H., and Hennon, P.E. 1985. Decline and mortality of Chamaecyparis nootkatensis in Southeastern Alaska, a problem of long duration but unknown cause. Plant Dis. 69:13-17.

419. Shaw, C.G. III, and Loopstra, E.M. 1988. Identification and pathogenicity of some Alaskan isolates of Armillaria. Phytopathology 78:971-974.
420. Sheldon, C. 1912. The wilderness of the North Pacific Coast Islands: a hunter's experience while searching for wapiti, bears, and caribou on the larger islands of British Columbia and Alaska. Scribner's Sons, New York. 246 pp.
421. Shigo, A.L. 1985. Compartmentalization of decay in trees. Sci. Am. 252:76-83.
422. Shortle, W.C., and Smith, K.T. 1988. Aluminum-induced, calcium deficiency syndrome in declining red spruce. Science 249:1017-1018.
423. Shriner, D.S., Heck, W.W., McLaughlin, S.B., Johnson, D.W., Joslin, J.D., and Peterson, C.E. 1989. Response of vegetation to atmospheric deposition and air pollution, NAPAP, SOS/T 18, Washington, DC.
424. Silverborg, S.B., and Brandt, R.W. 1957. Association of Cytophoma pruinosa with dying ash. For. Sci. 3:75-78.
425. Silverborg, S.B., and Ross, E.W. 1968. Ash dieback disease development in New York State. Plant Dis. Rep. 52:105-107.
426. Silvertown, J.W. 1982. Introduction to plant population ecology. Longman, Inc., New York. 209 pp.
427. Simons, E.E., Laudermilch, G.E., and Towers, B. 1983. Sugar maple dieback and mortality in northern Pennsylvania. Pennsylvania Bur. For., Div. Forest Pest Management, Middletown, PA. (Unpub. rep.)
428. Sinclair, W.A. 1965. Comparison of recent declines of white ash, oaks and sugar maple in northeastern woodlands. Cornell Plantations. 20:62-67.
429. Sinclair, W.A. 1967. Decline of hardwoods: possible causes. Proc. Inter. Shade Tree Conf. 42:17-32.
430. Sinclair, W.A., and Hudler, G.W. 1988. Tree declines: four concepts of causality. J. Arbor. 14:29-35.
431. Sinclair, W.A., Lyon, H.H., and Johnson, W.T. 1987. Diseases of trees and shrubs. Cornell Univ. Press. Ithaca, NY. 574 pp.
432. Sinner, K.F., and Rehfuess, K.E. 1972. Wirkungen einer Fomes annosus-Kernfäule auf den Ernährungszustand älterer Fichten (Picea abies Karst.). Allg. Forst Jagdztg. 143:74-80.

433. Sites, W.H. 1982. Inspection of white oaks. Memo to forest Supervisor, George Washington National Forest (3400); Harrisonburg, VA; July 26.
434. Skelly, J.M. 1974. Growth loss of scarlet oak due to oak decline in Virginia. Plant Dis. Rep. 58:396-399.
435. Skelly, J.M. 1987. A pathologist's view of the effects of changes in the chemical climate upon tree growth, Pages 161-170 in: Woody Plant Growth in a Changing Chemical and Physical Environment Proc. IUFRO Working Party S2.01-11. Shoot Growth Physiol., D.P. Lavender, ed. Vancouver, B.C., For. sci. Dept. Univ. B.C., 314 pp.
436. Skelly, J.M. 1989. Forest decline versus tree decline - the pathological considerations. Env. Mon. and Assess. 12:23-27.
437. Skelly, J.M., Davis, D.D., Merrill, W., Cameron, E.A., Brown, H.D., Drummond, D.B., and Dochinger, L.S. 1989. Diagnosing Injury to Eastern Forest Trees. Pennsylvania State University and USDA-Forest Service. PSU - Ag Mailing Room. Univ. Park, PA 122 pp.
438. Smith, H.E., and Keiser, G.M. 1971. Testing sugar maple sap for sweetness with a refractometer. USDA For. Serv. Res. Note NE-138. 4 pp.
439. Smith, J.B., and Tirpak, D. 1989. The potential effects of global climate change on the United States: Forests (Chapter 5). Report to Congress. United States Environmental Protection Agency. Washington, DC.
440. Solomon, A.M. 1986. Transient response of forests to CO_2-induced climate change: simulation modeling experiments in eastern North America. Oecologia 68:567-579.
441. Sperry, J.S., Donnelly, J.R., and Tyree, M.T. 1988. Seasonal occurrence of xylem embolism in sugar maple (Acer saccharum). Am. J. Bot. 75:1212-1218.
442. Staley, J.M. 1965. Decline and mortality of red and scarlet oaks. For. Sci. 11:2-17.
443. Stanosz, G.R. 1991. Symptoms, incidence, and pathogenicity of an anthracnose fungus from sugar maple seedlings in pear thrips-infested stands. Phytopathology 81:124.
444. Starkey, D.A., and Brown, H.D. 1986. Oak decline and mortality in the Southeast -- an assessment. Pages 103-114 in: Proc. 14th Ann. Hardwood Symp. of the

Hardwood Research Council; 18-21 May 9186; Cashiers, NC.

445. Starkey, D.A., Oak, S.W., Ryan, G.W., Tainter, F.H., Redmond, C., and Brown, H.D. 1989. Evaluation of oak decline areas in the south. USDA For. Serv. Protection Rep. R8-PR17. 36 pp.

446. Stemmermann, L. 1983. Ecological studies of Metrosideros in a successional context. Pac. Sci. 37:361-373.

447. Stemmermann, L. 1986. Ecological studies of 'ohi'a varieties (Metrosideros polymorpha Myrtaceae), the dominants in successional communities of Hawaiian rain forests. Ph.D. Diss. Univ. Hawaii. 200 pp.

448. Stephens, G.R., Turner, N.C., and DeRoo, H.C. 1972. Some effects of defoliation by gypsy moth (Porthetria dispar L.) and elm spanworm (Ennomos subsignarius Hbn.) on water balance and a growth of deciduous forest trees. For. Sci. 18:326-330.

449. Stipes, R.J., and Phipps, P.M. 1971. A species of Endothia associated with a canker disease of pin oak in Virginia. Plant Dis. Rep. 55:467-469.

450. Srayer, D.J., Glitzenstein, S., Jones, C.G., Kilasa, J., Likens, G.E., McDonnell, M.J., Parker, G.G., and Pickett, S.T.A. 1986. Long-term ecological studies: An illustrated account of their design, operation and importance to ecology. Inst. of Ecosystem Stud. Occ. Pub. 1.

451. Sucoff, E., Thornton, F.C., and Joslin, J.D. 1990. Sensitivity of the seedlings to aluminum: I. Honeylocust. J. Env. Qual. 19:163-171.

452. Swanson, R.H. 1983. Numerical and experimental analyses of implanted-probe heat pulse velocity theory. Ph.D. Thesis, Department of Botany, University of Alberta, Edmonton, Alberta, Canada.

453. Tainter, F.H., and Bensen, J.D. 1983. Effect of climate on growth, decline, and death of red oaks in western North Carolina. Phytopathology 72:838.

454. Tainter, F.H., Benson, D.M., and Fraedrich, S.W. 1984. The effect of climate on growth, decline, and death of the northern red oaks in the western North Carolina Nantahala Mountains. Castanea 49:127-137.

455. Tainter, F.H., Cody, J.B., and Williams, T.M. 1983. Drought as a cause of oak decline and death on the

South Carolina coast. Plant Dis. 67:195-197.
456. Tainter, F.H., Retzlaff, W.A., Starkey, D.A., and Oak, S.W. 1990. Decline of radial growth in red oaks is associated with short-term changes in climate. Eur. J. For. Path. 90:90-105.
458. Teck, R.M., and Hilt, D.E. 1990. Individual-tree probability of survival model for northeastern United States. USDA For. Serv. Res. Paper NE-642. 10 pp.
459. Tegethoff, A.C., and Brandt, R.W. 1964. Ash dieback disease development in New Hampshire, Vermont, Massachusetts, Connecticut, New Jersey and Pennsylvania, 1963. Plant Dis. Rep. 48:974-977.
460. Teillon, H.B, Burns, B.S., and Kelley, R.S. 1982. Forest insect and disease conditions in Vermont. VT Dep. For., Parks and Recreation. Montpelier. 28 pp.
461. Thériault, A., and Ansseau, C. 1990. Variabilité des protéines des samares d'érable rouge d'Amérique du Nord. Annales de l'ACFAS 58:50.
462. Thibault, M. 1989. Végétation et facteurs du milieu dans les régions écologiques du Québec méridional. Première partie: La zone feuillue. Rapport interne no. 313, Gouv. du Québec, Min. de l'Énergie et des Ressources, Québec. 392 pp.
463. Thimann, K.V. ed. 1989. Senescence in plants. CRC Press, Inc., Boca Raton, FL. 276 pp.
464. Roole, E.R., and Broadfoot, W.M. 1959. Sweetgum blight as related to alluvial soils of the Mississippi River floodplain. For. Sci. 5:2-9.
465. Tovar, D.C. 1989. Air pollution and forest decline near Mexico City. Env. Mon. and Assess. 12:49-58.
466. Towers, B., Simons, E.E., and Laudermilch, G.E. 1985. Sugar maple dieback - 1984 field season report. Pennsylvania Div. For. Pest Management, Bur. For., Middletown, PA.
467. True, R.P., and Tryon, E.H. 1965. Oak stem cankers initiated in the drought year 1953. Phytopathology 46:617-622.
468. Tryon, E.H., and True, R.P. 1958. Recent reductions in annual radial increments in dying scarlet oaks related to rainfall deficiencies. For. Sci. 4:219-230.
469. Tveite, B. 1987. Air Pollution and Forest Damage in Norway. Pages 59-67 in: NATO ASI Series, Vol. G16. Effects of Atmospheric Pollutants on Forests, Wetlands

and Agricultural Ecosystems. T.C. Hutchinson and K.M. Meema eds. Springer-Verlag, Berlin.
470. Tyree, M.T., and Sperry, J.S. 1989. Vulnerability of xylem to cavitation and embolism. Ann. Plant Physiol. Molecular Biol. 40:19-38.
471. Ulrich, B. 1980. Die Wälder in Mitteleuropa: Messergebnisse ihrer Umweltbelastung. Theorie einer Gefährdung. Prognose ihrer Entwicklung. Allg. Forst Z 35:1198-1202.
472. Ulrich, B. 1989a. Effects of acidic precipitation on forest ecosystems in Europe. Pages 189-272 in: Acidic Precipitation Vol. 2. Adriano, D.C. and Johnson, A.H. eds. Springer-Verlag, Berlin.
473. Ulrich, B. 1989b. Forest decline in ecosystem perspective. Pages 21-41 in: Proceedings International Congress on Forest Decline Research: State of Knowledge and Perspectives. Ulrich, B., ed. Kernforschungszentrum Karlsruhe GmbH. Postfach 3640, 7500 Karlsruhe 1.
474. Ulrich, B. 1990. Waldsterben: forest decline in West Germany. Environ. Sci. Technol. 24:436-441.
475. Van Deusen, P.C. 1990. Stand dynamics and red spruce decline. Can. J. For. Res. 20:743-749.
476. Van Valen, L. 1975. Life, death and energy of a tree. Biotropica 7:260-269.
477. Vieira, S.R., Hatfield, J.L., Nielsen, D.R., and Biggar, J.W. 1983. Geostatistical theory and application to variability of some agronomical properties. Hilgardia 51(3). 77 pp.
478. Villeneuve, J.P., Deschenes, S., Houle, S., Michaud, F., Jacques, G., and Grimard, Y. 189. Analyse de la variabilité spatiale des mesures de composition ionique des précipitations au Québec: Application de la méthode du krigeage aux données de précipitations acides. Rapport scientifique no. 205, INRS-Eau, Min. de l'Env. du Québec. 100 pages + annexes.
479. Vins, B. 1965. A method of smoke injury evaluation - determination of increment decrease. Pages 235-245 in: Commun. Inst. Forest. Cech., Praha.
480. Vogelmann, H.W. 1983. Acid rain: Are our forests in danger? Outdoor America 48(4):12-15.
481. Wagener, W.W. 1939. The canker of _Cupressus_ induced by _Coryneum cardinale_ n. sp. J. Agric. Res.

58:1-46.
482. Wagener, W.W. 1949. Top dying of conifers from sudden cold. J. For. 47:49-53.
483. Walker, S.L., and Auclair, A.N.D. 1989. Forest declines in western Canada and the adjacent United States. Federal LRTAP Liaison Office Report, Atmos. Environ. Serv., Dep. Environ., Downsview, Ontario, Canada. 150 pp.
484. Walker, S.L., Auclair, A.N.D., and Martin, H.C. 1991. Forest declines in eastern Canada. Federal LRTAP Liaison Office Report, Atmos. Environ. Serv., Dep. Environ., Downsview, Ontario, Canada.
485. Walker, T.W., and Syers, J.K. 1976. The fate of phosphorus during pedogenisis. Geoderma 15:1-19.
486. Wallace, H.R. 1978. The diagnosis of plant diseases of complex etiology. Ann. Rev. Phytopath. 16:379-402.
487. Wallace, W.R., and Hatch, A.B. 1953. Crown deterioration in the northern jarrah forests. Bound document, Forests Department, Western Australia, Australia.
488. Wargo, P.M. 1972. Defoliation-induced chemical changes in sugar maple roots stimulate growth of Armillaria mellea. Phytopathology 62:1278-1283.
489. Wargo, P.M. 1975. Lysis of the cell wall of Armillaria mellea by enzymes from forest trees. Physol. Plant Pathol. 5:99-105.
490. Wargo, P.M. 1976. Lysis of fungal pathogens by tree produced enzymes -- a possible disease resistance mechanism in trees. Pages 19-23 in: Proc. 23rd Northeastern Forest Tree Improvement Conference. Rutgers University, New Brunswick, NJ.
491. Wargo, P.M. 1977. Armillariella mellea and Agrilus bilineatus and mortality of defoliated oak trees. For. Sci. 23:485-492.
492. Wargo, P.M. 1978. Defoliation by the gypsy moth: how it hurts your tree. USDA Home and Gard. Bull. 223. 15 pp.
493. Wargo, P.M. 1980. Interaction of ethanol, glucose, phenolics and isolate of Armillaria mellea. Phytopathology 70:470.
494. Wargo, P.M. 1981. In vitro response to gallic acid of aggressive and non-aggressive "isolates" of Armillaria mellea. Phytopathology 71:565.

495. Wargo, P.M. 1983. The interaction of <u>Armillaria mellea</u> with phenolic compounds in the bark of roots of black oak. Phytopathology 73:838.
496. Wargo, P.M. 1988. Amino nitrogen and phenolic constituents of bark of American beech, <u>Fagus grandifolia</u>, and infestation by beech scale, <u>Cryptococcus fagisuga</u>. Eur. J. For. Pathol. 18:279-290.
497. Wargo, P.M., Bergdahl, D.R., Oleson, C.W., and Tobi, D.R. 1989. Root vitality and decline of red spruce. Report to the Forest Response Program, U.S. Environmental Protection agency, Corvallis, OR.
498. Wargo, P.M., and Houston, D.R. 1974. Infection of defoliates sugar maple trees by <u>Armillaria mellea</u>. Phytopathology 64:817-822.
499. Wargo, P.M., Parker, J., and Houston, D.R. 1972. Starch content in roots of defoliated sugar maple. For. Sci. 18:203-204.
500. Weiss, M.J., and Rizzo, D.M. 1987. Forest declines in major forest types of the eastern United States. Pages 297-306 in: Forest Decline and Reproduction: Regional and Global Consequences. Proc. of a Workshop held in Krakow, Poland, 23-27 March, 1987. K. Kairiukstis, S. Nilsson and A. Straszak, eds. Publ. No. WP-87-75. International Institute for Applied Systems Analysis, Laxenburg, Austria.
501. Wene, D.G. 1979. Stress predisposition of woody plants to <u>Botryosphaeria dothidea</u> stem canker. Urbana: University of Illinois. Ph.D. Thesis.
502. Wene, E.G., and Schoeneweiss, D.F. 1980. Localized freezing predisposition to <u>Botryosphaeria</u> canker in differentially frozen woody stems. Can. J. Bot. 58:1455-1459.
503. Werner, P.A., and Caswell, H. 1977. Population growth rates and age versus stage-distribution models for tcasel (<u>Dipsacus sylvestris</u> Huds.) Ecology 58:1103-1111.
504. Werner, W. 1988. Canopy dieback in the upper montane rain forest of Sri Lanka. GeoJournal 17:245-248.
505. Whittaker, R.H. 1954. The ecology of serpentine soil. IV. The vegetational response to serpentine soils. Ecology 35:275-288.
506. Wickman, B.E. 1986. Radial growth of grand fir and

Douglas-fir 10 years after defoliation by the Douglas-fir tussock moth in the Blue Mountains outbreak. USDA For. Serv. Res. Pap. PNW-367. 11 pp.

507. Widemann, E. 1924. Zuwachsrückgang und Wuchsstockungen der Fichte in den mittleren und unteren Höhenlagen der sächsischen Stattsforsten. 190 S. Akademische Buchhandlung Walter Laux, Tharandt i.S., FRG, pp. 190.

508. Widemann, E. 1927. Untersuchungen über das Tannensterben. Forstw. Centralbl. 49:759-780, 815-827, 845-853.

509. Wiegel, H., and Rabe, R. 1985. Wiederbesiderlung des Ruhrgebiets durch Flechten zeigt Verbesserung der Luftqualität an. Staub-Reinhaltung der Luft. 45:124-126.

510. Wigle, T.M.L. 1985. Impact of extreme events. Nature 316:106-107.

511. Wilkinson, R.C. 1987. Geographic variation in needle morphology of red spruce in relation to winter injury and declines. Pages 507-514 in: Proc. United States/Federal Republic of Germany Research Symposium; Effects of Atmospheric Pollutants on the Spruce-Fir Forests of the Eastern United States and the Federal Republic of Germany; 1987 October 19-23; Burlington, VT. USDA For. Serv. Gen. Tech. Rept. NE-120.

512. Willits, C.O. 1965. Maple syrup producers manual. USDA ARS, Agric. Handb. No. 134. 112 pp.

513. Wodzicki, J.T., and Brown, C.L. 1970. Role of xylem parenchyma in maintaining the water balance of trees. Acta Boc. Bot. Pol. 39:617.

514. Woodman, J.N., and Cowling, E.B. 1987. Airborne chemicals and forest health. Environ. Sci. and Tech. 21:120-126.

515. Woodwell, G.M. 1989. On causes of biotic impoverishment. Ecology 70:14-15.

516. Woolhouse, H.W. 1967. The nature of senescence in plants. Pages 179-214 in: Aspects of the Biology of Aging. Acad. Press Inc., New York. 634 pp.

517. Zak, B. 1961. Aeration and other soil factors affecting southern pines as related to littleleaf disease. USDA Tech. Bull. 1248. 30 pp.

518. Zech, W., und Popp, E. 1983. Magnesiummangel,

einer der Gründer für das Fichten und Tannensterben in NO-Bayern. Forstw. Centralbl. 102:50-55.

519. Zedaker, S.M., Eager, C., White, P.S., and Burk, T.E. 1988. Stand characteristics associated with potential decline of spruce-fir forests in the southern Appalachians. Pages 123-131 in: Proc. US/FRG Res. Symp. Effects of Atmospheric Pollutants on the Spruce-Fir Forests of Eastern United States and the Federal Republic of Germany. Burlington, VT. USDA-For. Serv. Gen'l. Tech. Rept. NE-120.

520. Zedaker, S.M., and Nicholas, N.S. 1988. Assessment of for est decline in the southern Appalachian spruce-fir forest, USA. Pages 239-244 in: Air Pollution and Forest Decline. J.B. Bucher and I. Bucher-Wallin, eds. Proc. 14th Int'l. Meeting for Specialists in Air Pollution Effects on Forest Ecosystems. IUFRO P2.05. Interlaken, Switzerland. Birmensdorf.

521. Zielonkowski, W., Preiss, H., and Heringer, J. 1986. Natur und Landschaft im Wandel. Berichte der Akademie für Naturschutz und Landschaftspflege, 8229 Laufen, Germany, 10:1-71.

522. Zimmerman, M.H. 1983. Xylem structure and the ascent of sap. Springer-Verlag, Berlin. New York.

523. Zöttl, H.W. 1990. Ernährung und dungung der ficte. Forstw. Cbl. 109:103-137.

524. Zöttl, H., and Hüttl, R.F. 1986. Nutrient supply and forest decline in Southwest Germany. Water, Air, Soil Pollut. 31:449-462.

525. Zöttl, H.W., and Hüttl, R.F. 1989. Nutrient deficiencies and forest decline, Pages 189-193 in: Air Pollution and Forest Decline, J.B> Bucher and I. Bucher-Wallin, eds. proc. 14th Int'l. Meeting for Specialists in Air Pollution Effects on Forest Ecosystems, IUFRO P2.05,Interlaken, Switzerland, Birmensdorf.

526. Zöttl, H.W., Hüttl, R.F., Fink, S., Tomlinson, G.H., and Wisniewski, J. 1989. Nutritional disturbances a n d historical changes in declining forests. Water, Air, and Soil Poll. 48:87-109.

INDEX

Abbau, 82
Abies:
  alba, 32, 59, 85
  balsamea, 51, 85
  Fraseri, 85
abiotic:
  agents, 87, 124, 169
  cause, 120
  disease, 26
  factor, 122, 182, 184
  stress, 20, 84
Acer:
  rubrum, 128
  saccharum, 12, 18, 39, 85, 123
acid:
  deposition, 23
  rain, 2, 76, 87, 88, 90, 95, 187, 189
  soil, 166
acidification, 81
acute yellowing, 76, 77, 79, 83
Adelges:
  picea, 94
  abietis, 91
aerial survey, 109
age-states, 28
aging, 35, 182, 184
Agrilus, 10
  anxius, 22
  bilineatus, 13, 16
Alabama, 168
Alaska, 39, 109, 109, 110
Alaska yellow-cedar, 39, 108, 186
alien tree species, 30
Alnus glutinosa decline, 83

altered tissues, 5
aluminum:
  calcium ratio, 21, 23, 76
  toxicity, 32, 76
American beech, 128, 146
amino acid, 9, 19
anaerobic soils, 114
annual:
  growth, 69
  mortality, 61, 188
  rings, 115
antagonistic component, 78
anthracnose, 18, 101
anthropogenic, 82, 85, 87, 106
Appalachian Mountains, 94
Armillaria sp., 9, 10, 12, 16, 22, 24, 57, 70, 118, 126, 169
ash, 39
  decline, 17
  dieback, 3, 4, 13, 15, 17, 47, 53
  rust, 13, 15
aspen, 144
Asterolecanium sp., 17
Asterosporium, 18
Austria, 64
auto-succession, 29, 34
autotoxicity, 33
awareness, 92
β-1,3 glucanase, 9
Baden-Wurttemberg, 98, 102
balsam fir, 51, 94
  dieback, 53
balsam wooly adelgid, 94, 95, 97
bark beetles, 119
bark moisture, 4, 17
basswood, 146
Bavaria, 61, 64, 65, 71, 72, 73, 74, 79, 80
Bay of Fundy, 42
bears, 119
beech, 62
  bark disease, 18
  death, 66
  leaf beetle, 101
Betula:

alba, 8
　<u>alleghaniensis</u>, 20, 38, 42
　<u>papyrifera</u>, 20, 42
biological invasion, 31
biophysical model, 57
biotic:
　agent, 34, 124, 169, 186
　disease, 26, 27, 65
　factors, 117, 182, 184
　pathogens, 86, 100
biotically impoverished, 34, 36
birch, 144
　dieback, 4, 20, 21, 42, 53
black ash, 42
black cherry, 128, 146
Black Forest, 64, 65, 72, 85, 91, 92
Black Mountains, 94
black spruce, 39
blight, 3
bog, 112, 120
bogus science, 107
bole lesions, 117
<u>Botryodiplodia</u>, 17
<u>Botryosphaeria</u> <u>dothidea</u>, 8, 10
British Colombia, 55, 109
bronze birch borer, 22
bruce spanworm, 147
Buchensterben, 66
California, 4, 22, 137
Canada, 12
　eastern, 17, 41, 138
canonical discriminant analysis, 169, 170, 172, 176, 179
canopy dieback, 28, 36
canopy dominants, 188
carbohydrates, 9
carbon dioxide, 68
case study, 127
catastrophic disturbance, 33
causality, 90
cavitation, 32, 47, 48, 49, 50, 52, 55
<u>Ceratocystis</u> <u>fagacearum</u>, 3
<u>Chamaecyparis</u> <u>nootkatensis</u>, 39
chemical injuries, 91

chestnut blight, 81, 182
China, 97
chitinase, 9
chlorosis, 139, 145
chronosequential monoculture, 30, 33
chronosequential polyculture, 30
Cibotium spp., 30
cicada, 101
climate, 21, 65, 74, 120, 121, 132, 156, 165
  change, 56, 169
  warming, 51
climatic:
  barriers, 31
  climax, 32
  events, 184
  extremes, 33
  fluctuations, 57
  instabilities, 33
  perturbation, 38
  zones, 148
climax, 189
cohort, 28, 29, 34
  senescence, 26, 33, 36
community ecology, 27
community dynamics, 29, 36, 189
competing species, 31
competition, 134, 188
competitive dominant, 188
complex diseases, 59
complex of diseases, 59, 84
conifers, 63
contour map, 158
contributing, 35, 122, 125, 126, 135, 137, 169, 184
correlation, 172
crown
  classes, 129
  condition, 178
  density, 183, 187, 188
  dieback, 43, 49, 50, 52, 123, 123, 125
  rating system, 137
  transparency, 60, 65, 66, 69, 72, 100, 186
Cryptococcus fagisuga, 18
Cylindrocarpon didymum, 118

cytokinin, 74
Cytophoma pruinosa, 17
Cytospora canker, 91
Czechoslovakia, 63
damage classes, 60, 62, 102
decline spiral, 27, 124, 184, 185
decline, 2, 3, 27, 40, 80, 81, 87, 89, 124, 155, 108, 183
  classes, 150
  concepts, 181
  definition, 184
  disease, 7, 26, 34, 132
  index, 139, 140, 141, 152, 187
  rating, 149
  sites, 170, 179
defoliation, 4, 5, 8, 12, 13, 15, 21, 105, 125, 126, 129, 138
demography, 35
Dendroctonus rufipennis, 24
dendrometer bands, 128
desiccation, 55, 121
destabilization, 36
deviation of the metabolism, 82
diagnosis, 85, 86, 184
Diaporthe alleghaniensis, 22
dieback, 3, 6, 27, 31, 40, 65, 87, 155, 159, 183
  classes, 129
  -decline disease model, 4
  distribution, 160
  index, 162, 167
  threshold, 51
discoloration, 188
disease, 59, 86, 125, 127, 109,
  concept, 181
displacement dieback, 30
dose-effect, 77
drainage, 112, 120
drought, 5, 8, 12, 16, 17, 21, 23, 32, 50, 52, 65, 67, 68, 88, 91, 95, 100, 125, 126, 132, 135, 169
Dutch elm disease, 81, 182
East Prussia, 65
ecological:
  cost, 182
  decline, 82
  gradient, 113

penalties, 188
   phase-shift, 82
ecosystem, 27, 31
ecosystemic complex diseases, 81
edaphic conditions, 31
edaphic extreme, 33, 34, 186
elevation, 67
embolism, 32
Endothia gyrosa, 10
energy reserves, 14
environmental pattern, 166
epidemic disease, 78
episodical outbreaks, 71
Eucalyptus marginata, 50
Europe, central, 32, 41, 63, 83, 87
European beech, 100, 101
Eutypella canker, 134
event analysis, 41
evolutionary:
   adaptation, 32
   fit, 33
   stress, 34
excessive moisture, 114
extreme winter cold, 21
extremes, 51
facultative organisms, 4
Fagus:
   grandifolia, 18, 128
   sylvatica, 18, 101
false spring, 47
Fichtensterben, 65, 66
field trials, 143
fir death, 66
fire, 29, 30, 109
foliage discoloration, 60, 186
forest:
   condition classes, 98, 99
   decline, 59, 85, 89, 95, 105, 181, 190
   dieback, 38, 39
   health survey, 60, 137
   management, 68, 81
   relationships, 11
   species decline, 184

tent caterpillar, 126, 129, 134, 147
France, 68
Fraser fir, 85, 94, 95
Fraxinus:
  americana, 13, 17, 38
  nigra, 42
  pennsylvanica, 17
freezing, 8, 21, 23, 41, 44, 50, 55, 91, 121, 169
frost, 32, 44, 47, 50, 125, 126, 135
  desiccation, 47, 48, 49, 50
fruit trees, 83
fungi, 117
fungicides, 78
Fusicoccum sp., 17, 18
Ganoderma root rot, 169
gap dynamics, 36
generalistic colonizer, 31
genetics, 184
geographic information systems, 150, 152, 166
Georgia, 168
geostatistics, 157, 166
germ theory, 1, 181
global warming, 39, 51, 189
global climate change, 52, 54
Glycobius speciosus, 126
Gnomonia spp. 101
greenhouse effect, 189
greenhouse climate, 36
ground survey, 111
growth, 131
  reductions, 80
  stagnation, 34
gypsy moth, 13, 147, 169
hail, 100
hardpan, 52, 90
hardwood decline, 81
Hawaii, 30
hazard rating, 168, 169
healthy, 128, 143, 151, 169, 187
heart rot, 73
heat disease, 65
heat injury, 91
hemispheric warming, 51

herbicides, 91
Heterobasidion annosum, 70
Heterocampa guttivitta, 12, 127
highgraded, 135, 136
historical:
  observations, 116
  evidence, 38
  examples, 124, 125, 187
host-stress-saprogen model, 6
Hypoxylon atropunctatum, 16
Hypoxylon canker, 169
inciting, 34, 35, 78, 122, 125, 126, 135, 137, 184
increment measurements, 63, 67
increment reductions, 139
injuries, 134
inoculations, 117, 118
insect, 86, 100, 109, 127
insecticide, 98
iron toxicity, 32
ironwood, 146
isolations, 117
Italy, 70
jarrah, 50
Koch's postulates, 18
kriging, 157, 158, 164
late frost, 4
Laurentian Mountains, 42
lead species, 43
leaf area index, 66, 69
leaf loss, 156
Lepidosaphes ulmi, 18
lesions, 117
Leucostoma kunzei, 10
lichens, 63
life stages, 28
lightning, 102, 104
little-leaf disease, 4, 22, 90
loblolly pine, 4, 22
longevity, 109
low temperature, 184
Lymantria dispar, 13
magnesium, 70, 76, 90, 91
Malacosoma disstria, 12, 126

management, 125, 169, 181, 186, 187
manganese toxicity, 32
maple:
  blight, 4, 12, 126
  decline, 4, 155
  dieback, 4, 53, 155, 187
  leafcutter, 134
  sap, 131
  syrup, 141
  webworm, 12, 134
maturity, 182
Metrosideros polymorpha, 30
Mexico, 97
mineral deficiencies, 91
Mississippi, 168
Mittelgebirge, Germany, 76
MLO, 3, 17, 20, 82, 83
model, host-stress-saprogen, 3
monocarpic dieback, 28
mortality, 95, 108, 147, 156, 188
Mount Mitchell, 95, 96, 98
mountain races, 66
mountain hemlock, 109
mountainous yellowing, 76
mulberry, 82
Mycelium radicis atrovirens, 118
mycorrhizae, 78, 119
natural:
  decline, 122
  dieback, 33, 35
  disturbance, 127
  forest, 10
  perturbations, 88
  phenomenon, 27, 30, 31
  populations, 181
  process, 36
  regression, 32
  selection, 188, 189
Nectria, 18
  cinnabarina, 10
  coccinea var. *faginata*, 19
  galligena, 19
  ochroleuca, 19

nematodes, 119
neuartig Waldschäden, 89
New Brunswick, 42
New England, 12, 20, 127
new forest decline, 54
New Hampshire, 20
new-type forest damage, 89
New York, 12, 20, 127, 132
nitrogen, 68
  oxide, 63, 64
normal, 70, 86, 183
North Carolina, 16, 94
northern hardwoods, 40, 41
Norway maple, 187
Norway spruce, 29, 38, 41, 54, 55, 59, 62, 65, 66, 67, 70, 73, 76, 77, 78, 79, 83, 85, 90, 91, 92, 137
Nova Scotia, 20, 42
novel forest damage, 60, 72, 76, 81, 83
novel forest disease, 29
nugget, 157, 160
nutrient:
  deficiencies, 89
  disorder, 100
  limitations, 33
oak, 26, 39
  decline, 4, 13, 16, 21, 168, 169, 178
Oligonychus ununguis, 89
onset, 43, 115
Ontario, 20, 42, 137
opportunistic organisms, 4, 7
ordination, 112
Ore Mountains, 63
Oregon, 109
organic matter, 178
organisms of secondary action, 5, 184
overstocked, 136
ozone, 63, 64, 76, 88, 106
Pacific Northwest, 54
Pacific Rim, 52
paludification, 120
parasites, 26
patch dynamics, 36
pathogen, 118, 122

patterns in dieback, 156
Perenniporia sabacida, 24
periodically recurring perturbations, 33
perturbations, 32, 34, 186
pH, 63, 64, 178
phenolics, 19
Phloeosinus spp. 119
Phomopsis, 18, 22
phosphorus deficiency 166
photographs, 66, 88, 115
photooxidants, 67, 76
physiological ecology, 35
physiological shock, 32, 34
Phytophthora, 57
   gonapodyides, 119
Picea:
   abies, 23, 29, 38, 59, 85
   mariana, 39
   rubens, 23, 38, 85
   sitchensis, 109
Piedmont region, 4
pines, 59
Pinus:
   contorta var. contorta, 109
   echinata, 4, 22
   monticola, 4, 22, 41
   ponderosa, 4, 22
   taeda, 4, 22
plant communities, 112
pole blight, 54, 55
pollutant, 23, 39, 56, 60, 62, 68, 70, 72, 74, 76, 81, 84, 85, 87, 88, 90, 95, 99, 101, 105, 122, 123, 137, 154, 182, 190
polycarpic trees, 35
ponderosa pine, 4, 22
poor soils, 169
poorly drained soil, 148, 186
population ecology, 27
Populus tremuloides, 50
postcards, 66
potato decline, 82
powdery mildew, 101
power plants, 63
predispose, 7, 14, 18, 34, 36, 81, 125, 126, 135, 184

primary cause, 122
productivity stress, 82
Prunus serotina, 128
Puccinia sparganioides, 13
purturbation pulse, 34
Pyrenochaeta, 18
Pythium, 119
quality assurance, 138, 145
Quebec, 20, 21, 39, 41, 137, 155
Quercus:
  coccinea, 90
  falcata, 16
  laurifolia, 16
  nigra, 16
  phellos, 16
  prinus, 17
  rubra, 16, 90
radial growth decrease, 40
radiation, geographic, 31
radiation, adaptive, 31
recolonize, 63
recovery, 5, 43, 67, 72, 89, 92, 125, 132, 135, 139, 149, 183, 189
red:
  oak, 90, 146, 173
  maple, 128, 146, 156
  spruce, 38, 94, 95, 98
    decline, 21, 23
redox shift, 82
refoliate, 14, 147
regeneration, 114
regreening, 76, 77, 79
regressive soil, 32
release, 115
remote sensing, 157
replacement dieback, 30
reproduction, 114
Rheinland-Pflanz, 64
rhizosphere, 78
Rhynchaenus fagi, 101
Rickettsia, 82
rime ice, 89, 95, 100
ring analysis, 67

root:
  disease, 78
  freezing, 12, 55
  lesions, 117
  pathogens, 75
  tyloses, 52
rope trees, 115
Ruhr, 63, 64
running out, 82
saddled prominent, 12, 127, 129
salt, 12, 91
salt-and-pepper dieback, 36
salvage, 61
San Bernardino Mountains, 4, 22
San Gabrial Mountains, 4, 22
saprobe, 118
saprogen, 4, 9, 10, 14
sapstreak disease, 130, 134
Saxony, 71
scales, 17, 18
scarlet oak, 90
scars, 119
Schleswig-Holstein, 61, 64
<u>Scytinostroma galactinum</u>, 24
secondary, 122
  organisms, 4, 5, 9
seed production, 129
seedlings, 31, 114
<u>Seiridium cardinale</u>, 118
self thinning, 188
senescence, 35, 182, 184
senescence, premature, 100
senescing cohort, 29
shore pine, 109
shortleaf pine, 4, 22
sill, 157, 160
silver fir, 19, 32, 59, 70, 71, 74, 75, 85, 92, 93, 94, 102, 103, 104, 137
silvicultural treatment, 135, 180, 187
simplified forest structure, 33, 184
Sitca spruce, 109
site, 169
  index, 163, 164, 165

quality, 155, 156, 157, 159, 164, 187
snag classes, 111, 115
snow, 12, 121
  pack, 41, 47
soil:
  enrichment, 32
  factors, 148
  frost, 47, 50
  moisture, 126
  temperature, 20
  toxicity, 120
South Carolina, 16
spatial distribution, 157
spatial structure, 156, 160, 164, 166
species decline, 90
Sphagnum spp., 120
spider mite, 89
spittlebugs, 17
spread, 113
spruce gall adelgid, 91
stabilizing, 183
stand characteristics, 179
stand density, 126
stand-level dieback, 31, 35
stand-reduction dieback, 30, 32
starch, 9
Stegonsporium ovatum, 12
stem rot, 70
stork's nest, 71
stress-altered tissues, 5
stress tolerant dominants, 188, 190
succession, 30, 36, 109, 189
sucking insects, 4, 17
sugar content, 130
sugar maple, 13, 18, 39, 41, 49, 50, 85, 146, 156
  borer, 126, 132, 134
  decline, 12, 21, 123, 137, 139, 186
  dieback, 42
sugarbush, 124, 128
sulfur, 178
  dioxide, 63, 64
surveys, 96, 100, 102, 105, 154, 183, 186
sweetgum blight, 4

Switzerland, 61, 66, 68, 69, 72
symptoms, 117, 139, 169
synchronizing, 183
Taeniothrips inconsequens, 18
Tannensterben, 66, 70, 71, 72, 73, 75, 80
temperature increase, 4
Tennesee, 168
terminology, 88
Tetralopha asperatella, 12
thaw-freeze, 41, 47, 51, 54
Thiessen, polygon, 150
thinned, 136, 135
thrips, 18
Thuja plicata, 111
Tilia americana, 18
timber production, 68
topographic variables, 178
treatment, 169
tree growth, 164
treefall gaps, 30
trembling aspen, 50
triggering stress, 6
Tsuga mertensiana, 109
Tsuga heterophylla, 109
twolined chestnut borer, 13, 169
Type II error, 182
Typhlocyba cruenta, 101
unfavorable drainage, 156
United States:
  eastern, 23, 51
  northeast, 17
  southern, 168
unsettled tree diseases, 80
vapor block, 50
variogram, 157, 160
Vermont, 90, 126
viroid, 82
virus, 17, 21, 82, 83
Waldsterben, 29, 54, 56, 59, 60, 65, 67, 68, 76, 81, 83, 89, 90, 92, 181
water:
  potential, 8
  shortage, 4

stress, 8, 49
table, 21
waterlogging, 50, 120
weak organisms, 4
weather, 94, 98, 145
  disturbance, 33
weighting factor, 143
West Germany, 55, 59, 87, 89, 98, 99, 137, 181, 186
western hemlock, 109, 113
western redcedar, 111, 112
white:
  ash, 15, 42, 38, 146
  birch, 20, 146
  pine, western, 4, 22, 41, 54, 55
wind, 23, 100
winter frost resistance, 41
winter thaw, 47, 132, 135
Wisconsin, 12, 126
wounds, 102
xylem injury, 48
yellow birch, 20, 38, 50, 146
yellowing, 70, 76, 100, 105
zero force, 189